50 CHALLENGING ALGEBRA PROBLEMS

$3x - 2y$

$9x^2 - 12xy + 4y^2$

$27x^3 - 54x^2y + 36xy^2 - 8y^3$

FULLY SOLVED

Chris McMullen, Ph.D.

50 Challenging Algebra Problems (Fully Solved)
Improve Your Math Fluency
Chris McMullen, Ph.D.

www.improveyourmathfluency.com

Zishka Publishing
ISBN: 978-1-941691-23-6

Textbooks > Math > Algebra
Study Guides > Workbooks> Math
Education > Math > Algebra

Problem 1

Directions: Solve for x in the equation below. (Don't use guess and check.)

$$\sqrt{x} + \sqrt{2} = \sqrt{32}$$

❖ You can find the solution on the following page.

Solution to Problem 1

Subtract $\sqrt{2}$ from both sides of the equation in order to isolate the unknown term.
$$\sqrt{x} = \sqrt{32} - \sqrt{2}$$
Square both sides of the equation to get the variable out of the squareroot. Apply the rules $\left(\sqrt{x}\right)^2 = x$ and $(a - b)^2 = a^2 - 2ab + b^2$ where $a = \sqrt{32}$ and $b = \sqrt{2}$.
$$\left(\sqrt{x}\right)^2 = \left(\sqrt{32} - \sqrt{2}\right)^2$$
$$x = \left(\sqrt{32}\right)^2 - 2\sqrt{32}\sqrt{2} + \left(\sqrt{2}\right)^2$$
Apply the rules $\left(\sqrt{a}\right)^2 = a$ and $\sqrt{a}\sqrt{b} = \sqrt{ab}$.
$$x = 32 - 2\sqrt{(32)(2)} + 2$$
Note that $32 + 2 = 34$ and $(32)(2) = 64$.
$$x = 34 - 2\sqrt{64}$$
Note that $\sqrt{64} = 8$.
$$x = 34 - (2)(8)$$
$$x = 34 - 16$$
$$x = 18$$

Check the answer: Plug $x = 18$ into the original equation.
$$\sqrt{x} + \sqrt{2} = \sqrt{32}$$
$$\sqrt{18} + \sqrt{2} = \sqrt{32}$$
Use a calculator.
$$4.242640687 + 1.414213562 = 5.656854249$$
The answer checks out.

Common mistake: It's incorrect to rewrite $\sqrt{x} + \sqrt{2} = \sqrt{32}$ as $x + 2 = 32$. You can easily verify that x doesn't equal 30 by plugging 30 in for x in the original equation with a calculator. You can't square each term individually in algebra. You can square both sides of the equation, but then you get a cross term when you f.o.i.l. it out:
$$\left(\sqrt{x} + \sqrt{2}\right)^2 = \left(\sqrt{32}\right)^2$$
$$\left(\sqrt{x}\right)^2 + 2\sqrt{x}\sqrt{2} + \left(\sqrt{2}\right)^2 = \left(\sqrt{32}\right)^2$$
That's why $x + 2 = 32$ is incorrect. It should really be $x + 2\sqrt{x}\sqrt{2} + 2 = 32$.

Problem 2

Directions: Solve for x and y in the system of equations below.

$$x + y = 16$$

$$\frac{1}{x} + \frac{1}{y} = \frac{1}{3}$$

❖ You can find the solution on the following page.

Solution to Problem 2

Subtract y from both sides of the first equation.
$$x = 16 - y$$
Substitute $16 - y$ in place of x in the second equation.
$$\frac{1}{16 - y} + \frac{1}{y} = \frac{1}{3}$$
Add the left fractions by first making a common denominator. Multiply $\frac{1}{16-y}$ by $\frac{y}{y}$ and multiply $\frac{1}{y}$ by $\frac{16-y}{16-y}$ in order to make a common denominator.
$$\frac{1}{16 - y}\left(\frac{y}{y}\right) + \frac{1}{y}\left(\frac{16 - y}{16 - y}\right) = \frac{1}{3}$$
$$\frac{y}{16y - y^2} + \frac{16 - y}{16y - y^2} = \frac{1}{3}$$
To add fractions that share a common denominator, add their numerators.
$$\frac{y + 16 - y}{16y - y^2} = \frac{1}{3}$$
$$\frac{16}{16y - y^2} = \frac{1}{3}$$
Cross multiply. Recall that $\frac{w}{x} = \frac{y}{z}$ becomes $wz = xy$ when you cross multiply.
$$(16)(3) = (1)(16y - y^2)$$
$$48 = 16y - y^2$$
This is a quadratic equation. First express the equation in standard form.
$$y^2 - 16y + 48 = 0$$
Either factor this equation or apply the quadratic formula. We will factor it.
$$(y - 4)(y - 12) = 0$$
The two solutions are $y = 4$ and $y = 12$. Plug each solution into $x = 16 - y$. The two corresponding solutions are $x = 12$ and $x = 4$.

Check the answers: Plug $x = 4$ and $y = 12$ into the original equations.
$$x + y = 4 + 12 = 16$$
$$\frac{1}{x} + \frac{1}{y} = \frac{1}{4} + \frac{1}{12} = \frac{3}{12} + \frac{1}{12} = \frac{4}{12} = \frac{1}{3}$$
You will get the same thing using $x = 12$ and $y = 4$. The answers check out.

Problem 3

Directions: Solve for x in the equation below, where x is a positive real number.

$$2x^{4/3} + 5x^{3/4} = 6x^{4/3} - 3x^{3/4}$$

❖ You can find the solution on the following page.

Solution to Problem 3

Combine like terms: Add $3x^{3/4}$ to both sides and subtract $2x^{4/3}$ from both sides.

$$5x^{3/4} + 3x^{3/4} = 6x^{4/3} - 2x^{4/3}$$

Note that $5x^{3/4} + 3x^{3/4} = 8x^{3/4}$ and $6x^{4/3} - 2x^{4/3} = 4x^{4/3}$. One way to see this is to factor: For example, $5x^{3/4} + 3x^{3/4} = (5+3)x^{3/4} = 8x^{3/4}$.

$$8x^{3/4} = 4x^{4/3}$$

Divide both sides of the equation by 4.

$$2x^{3/4} = x^{4/3}$$

Divide both sides of the equation by $x^{3/4}$. (The problem states that x is positive, so we don't need to worry about dividing by zero.) Note that $\frac{x^{3/4}}{x^{3/4}} = 1$.

$$2 = \frac{x^{4/3}}{x^{3/4}}$$

Apply the rule $\frac{x^m}{x^n} = x^{m-n}$.

$$2 = x^{4/3 - 3/4}$$

Subtract fractions by making a common denominator. Focus on the exponent.

$$\frac{4}{3} - \frac{3}{4} = \frac{4}{3}\left(\frac{4}{4}\right) - \frac{3}{4}\left(\frac{3}{3}\right) = \frac{16}{12} - \frac{9}{12} = \frac{7}{12}$$

Since $4/3 - 3/4$ equals $7/12$, the previous equation can be expressed as:

$$2 = x^{7/12}$$

Raise both sides of the equation to the power of $12/7$. Why? We will apply the rule that $(x^m)^n = x^{mn}$ such that $(x^m)^{\frac{1}{m}} = x^{m\left(\frac{1}{m}\right)} = x^1 = x$.

$$(2)^{\frac{12}{7}} = \left(x^{\frac{7}{12}}\right)^{\frac{12}{7}}$$

$$(2)^{\frac{12}{7}} = x$$

If you enter 2^(12/7) on a calculator, you will find that $x \approx 3.281341424$.

Check the answer: Plug $x \approx 3.281$ into the original equation. Use a calculator.

$$2x^{4/3} + 5x^{3/4} = 6x^{4/3} - 3x^{3/4}$$

$$2(3.281)^{4/3} + 5(3.281)^{3/4} \approx 6(3.281)^{4/3} - 3(3.281)^{3/4}$$

$$9.751 + 12.189 \approx 29.252 - 7.314$$

Since this works out to $21.940 \approx 21.938$, the answer checks out.

Problem 4

Directions: Determine the value of x in the equation below.

$$x^x = 256$$

❖ You can find the solution on the following page.

Solution to Problem 4

Solving for x by applying algebraic operations to both sides of this equation would be challenging. However, it turns out that this problem is numerically simple enough to figure out if you understand what the problem means conceptually.

The problem, $x^x = 256$, is equivalent to asking, "What number can you raise to the power of itself and obtain 256 as the result?"

Let's try raising integers to the powers of themselves and see what happens.
- $1^1 = 1$
- $2^2 = 2 \times 2 = 4$
- $3^3 = 3 \times 3 \times 3 = 27$
- $4^4 = 4 \times 4 \times 4 \times 4 = 256$

Hey, that happens to be the answer: $4^4 = 256$. This shows that $x = 4$ solves the equation $x^x = 256$.

Check the answer: Plug $x = 4$ into the original equation.
$$x^x = 256$$
$$4^4 = 256$$
Enter 4^4 on a calculator to see that 4^4 = 256. The answer checks out.

Problem 5

Directions: Determine the equation of a straight line that has a y-intercept of 1 and which is perpendicular to the line that is represented by the equation below.

$$y = 2x + 3$$

❖ You can find the solution on the following page.

Solution to Problem 5

Compare the given equation, $y = 2x + 3$, to the standard equation for a straight line, which is $y = mx + b$. The given line has a slope of $m = 2$ and a y-intercept of 3.

We wish to find a new line that is perpendicular to the given line. Recall that lines are perpendicular if one slope is the negative of the reciprocal of the other. That is,

$$m_\perp = -\frac{1}{m}$$

where $m_\perp$ is the slope of the perpendicular line. Plug in $m = 2$.

$$m_\perp = -\frac{1}{2}$$

The perpendicular line has a slope of negative one-half. According to the problem, the perpendicular line needs to have a y-intercept of 1, which we can express as

$$b_\perp = 1$$

Plug the slope $\left(m_\perp = -\frac{1}{2}\right)$ and y-intercept ($b_\perp = 1$) into the general equation for a straight line in order to write an equation for the perpendicular line.

$$y = m_\perp x + b_\perp$$
$$y = -\frac{x}{2} + 1$$

Check the answer: Here is a trick that you may learn in a trigonometry class that can help you check the answer to questions like this using a calculator. (Be sure that the calculator is in degrees mode and not in radians mode.) Take the inverse tangent of the slope, $\tan^{-1}(m)$, in order to determine the angle that the line makes with the x-axis. Be sure to use a $\tan^{-1}$ button or an arctangent (atan) button. Don't take $\frac{1}{x}$ of the tangent (because that would give you cotangent instead of the inverse tangent: in trigonometry, $\tan^{-1}$ doesn't mean one over the tangent). According to a calculator, for the given line, $\tan^{-1}(m) = \tan^{-1}(2) \approx 63.435°$, while for the perpendicular line, $\tan^{-1}(m_\perp) = \tan^{-1}\left(-\frac{1}{2}\right) \approx -26.565°$. One line is angled 63.435° above the x-axis, while the other line is angled 26.565° below the x-axis (since the angle is negative). Add these two angles to see that they make a right angle: $63.435° + 26.565° = 90°$. If you have a graphing calculator, graph each line.

Problem 6

Directions: Apply algebra to solve the following word problem.

A customer has ten coupons. Each coupon offers a 10% discount off of the purchase price. Unlike many store coupons, these coupons don't come with any restrictions (meaning that the customer may apply all ten coupons to the same purchase). The customer buys a pair of boots with a retail price of $50. How much will the customer pay if the customer uses all ten coupons and there is a sales tax of 8%? (Why aren't the boots free?)

❖ You can find the solution on the following page.

Solution to Problem 6

Before we apply algebra, let's reason out why the boots aren't free. It's incorrect to multiply the number of coupons (10) by the percentage (10%). Why? Stores apply coupons one at a time, not all together.

Here is what the program in the cash register does. The retail price of the boots is $50. When the first coupon is applied, the program deducts 10% from the purchase price, subtracting $5 from $50 to make a new subtotal of $45. When the second coupon is applied, the program deducts 10% off the new purchase price, subtracting $4.50 from $45 to make a new subtotal of $40.50. Although the first coupon was worth a $5 discount, the second coupon was worth a $4.50 discount because the new purchase price was lower. Each coupon results in a smaller discount than the previous discount as the purchase price is reduced. That's why the boots aren't free.

To carry out the algebra, we will multiply the retail price ($50) of the boots by 0.9 ten times. Instead of subtracting 10% from the purchase price ten times (which would require determining the new purchase price after each discount), we will multiply $50 by 90% ten times (since multiplying by 90% is the same as reducing the price by 10%, as $100\% - 10\% = 90\%$). In decimal form, 90% equals 0.9 (divide by 100 to determine this). Multiplying by 0.9 ten times equates to raising 0.9 to the power of 10. Finally, we will multiply by 1.08 to add the 8% sales tax (add 8% to 100% to get 108%, then divide by 100 to turn 108% into a decimal). The customer will pay a total amount given by:

$$x = (\$50)(0.9)^{10}(1.08)$$

Enter 50×1.08×.9^10 on a calculator to determine that the total cost is $18.83.

Check the answer: Deduct 10% ten times, and then add 8% sales tax. (The problem says to apply algebra in the solution, but we're just checking the answer now.)
(1) $50 - \$5 = \$45$ (2) $45 - \$4.50 = \$40.50$ (3) $40.50 - \$4.05 = \$36.45$
(4) $36.45 - \$3.65 = \$32.80$ (5) $32.80 - \$3.28 = \$29.52$ (6) $29.52 - \$2.95 = \$26.57$
(7) $26.57 - \$2.66 = \$23.91$ (8) $23.91 - \$2.39 = \$21.52$ (9) $21.52 - \$2.15 = \$19.37$
(10) $19.37 - \$1.94 = \$17.43$ (tax) $17.43 + \$1.39 = \$18.82$
The answer checks out (it's off by one penny due to rounding).

Problem 7

Directions: Find the real solutions for x in the equation below.

$$12x^2 = 63 - 3x^4$$

❖ You can find the solution on the following page.

Solution to Problem 7

This is actually a quadratic equation. One way to see this is to let $y = x^2$.
$$12x^2 = 63 - 3x^4$$
$$12y = 63 - 3y^2$$
Add $3y^2$ to both sides and subtract 63 from both sides to put this in standard form.
$$3y^2 + 12y - 63 = 0$$
It's convenient to divide both sides of the equation by 3.
$$y^2 + 4y - 21 = 0$$
Compare this to the general equation $ay^2 + by + c = 0$ to determine that:
$$a = 1 \quad , \quad b = 4 \quad , \quad c = -21$$
Plug these values into the quadratic formula. (Alternatively, you could factor the previous quadratic equation.)
$$y = \frac{-b \pm \sqrt{b^2 - 4ac}}{2a} = \frac{-4 \pm \sqrt{4^2 - 4(1)(-21)}}{2(1)} = \frac{-4 \pm \sqrt{16 + 84}}{2}$$
Note that the two minus signs make a plus sign: $-4(1)(-21) = (-4)(-21) = +84$.
$$y = \frac{-4 \pm \sqrt{100}}{2} = \frac{-4 \pm 10}{2}$$
There are two possible answers for y:
$$y = \frac{-4 + 10}{2} = \frac{6}{2} = 3 \quad \text{or} \quad y = \frac{-4 - 10}{2} = -\frac{14}{2} = -7$$
We're not finished yet because we need to solve for x. Recall that $y = x^2$. Squareroot both sides to get $\pm\sqrt{y} = x$. Why $\pm$? That's because $\left(\sqrt{y}\right)^2 = y$ and $\left(-\sqrt{y}\right)^2 = y$. Only $y = 3$ gives a real solution for x (since $\sqrt{-7}$ is imaginary). The two real answers are:
$$x = \pm\sqrt{3}$$

Check the answers: Plug $x = \pm\sqrt{3}$ into the original equation.
$$12x^2 = 63 - 3x^4$$
$$12\left(\pm\sqrt{3}\right)^2 = 63 - 3\left(\pm\sqrt{3}\right)^4$$
Note that $\left(\pm\sqrt{3}\right)^2 = \sqrt{3}\sqrt{3} = 3$ and $\left(\pm\sqrt{3}\right)^4 = (3)(3) = 9$.
$$12(3) = 63 - 3(9) = 36$$
The answers check out.

Problem 8

Directions: Find positive solutions for x, y, and z in the system of equations below.

$$x = yz$$

$$y = z(x + 1)$$

$$z = (x - 1)y$$

❖ You can find the solution on the following page.

Solution to Problem 8

Substitute the top equation ($x = yz$) into the bottom equations.
$$y = z(yz + 1) \quad \text{and} \quad z = (yz - 1)y$$
Multiply the left-hand sides together and the right-hand sides together.
$$yz = z(yz + 1)(yz - 1)y$$
Apply the f.o.i.l. method: $(a + b)(c + d) = ac + ad + bc + bd$.
$$yz = yz(y^2z^2 - yz + yz - 1)$$
$$yz = yz(y^2z^2 - 1)$$
Divide both sides of the equation by yz (we don't need to worry about dividing by zero since the problem specified positive solutions). Note that $\frac{yz}{yz} = 1$.
$$1 = y^2z^2 - 1$$
Add 1 to both sides of the equation to get $2 = y^2z^2$. Squareroot both sides to get $\sqrt{2} = yz$. Divide both sides of the equation by z to get $y = \frac{\sqrt{2}}{z}$. Plug this expression for y into the second equation at the top of this page.
$$z = \left(\frac{\sqrt{2}}{z}z - 1\right)\left(\frac{\sqrt{2}}{z}\right) = (\sqrt{2} - 1)\left(\frac{\sqrt{2}}{z}\right) = \frac{(\sqrt{2} - 1)\sqrt{2}}{z} = \frac{\sqrt{2}\sqrt{2} - 1\sqrt{2}}{z} = \frac{2 - \sqrt{2}}{z}$$
Multiply both sides of the equation by z to get $z^2 = 2 - \sqrt{2}$. Squareroot both sides.
$$z = \sqrt{2 - \sqrt{2}} \approx 0.76537$$
Plug this expression into the equation $y = \frac{\sqrt{2}}{z}$.
$$y = \frac{\sqrt{2}}{\sqrt{2 - \sqrt{2}}} \approx 1.84776 \left(\text{this is the same as } y = \sqrt{2 + \sqrt{2}} \approx 1.84776\right)$$
Plug y and z into the equation $x = yz$. The $\sqrt{2 - \sqrt{2}}$'s cancel out.
$$x = \left(\frac{\sqrt{2}}{\sqrt{2 - \sqrt{2}}}\right)\left(\sqrt{2 - \sqrt{2}}\right) = \sqrt{2} \approx 1.41421$$

Check the answers: Plug x, y, and z into the original equations.
$$x = yz \approx (1.84776)(0.76537) \approx 1.41422$$
$$y = z(x + 1) \approx 0.76537(1.41421 + 1) \approx 1.84776$$
$$z = (x - 1)y \approx (1.41421 - 1)1.84776 \approx 0.76536$$
The answers check out (within reasonable rounding tolerance).

Problem 9

Directions: Solve for x in the equation below. (Don't use guess and check.)

$$x\sqrt{x} = 27$$

❖ You can find the solution on the following page.

Solution to Problem 9

Compare the wording of this problem to the wording of Problem 4. Notice that this problem specifically states to "solve for x." Unlike Problem 4, guess and check would not be satisfactory based on this difference in the instructions. (Most teachers don't give full credit for guess and check solutions when known techniques could be used to solve for an answer, regardless of the wording.)

Recall that a squareroot can be expressed as a fractional exponent.

$$\sqrt{x} = x^{1/2}$$

Substitute $x^{1/2}$ in place of $\sqrt{x}$ in the given equation.

$$xx^{1/2} = 27$$

Note that $x^1 = x$. Plug this into the previous equation.

$$x^1 x^{1/2} = 27$$

Apply the rule that $x^m x^n = x^{m+n}$.

$$x^{1+1/2} = 27$$

Note that $1 + 1/2 = 1 + 0.5 = 1.5 = 3/2$ (or make a common denominator to write $1 + 1/2 = 2/2 + 1/2 = 3/2$).

$$x^{3/2} = 27$$

Raise both sides of the equation to the power of 2/3. Why? We will apply the rule that $(x^m)^n = x^{mn}$ such that $(x^m)^{\frac{1}{m}} = x^{m\left(\frac{1}{m}\right)} = x^1 = x$.

$$\left(x^{\frac{3}{2}}\right)^{\frac{2}{3}} = (27)^{2/3}$$

$$x = (27)^{2/3}$$

The power of 2/3 means to square 27 and take the cube root, in any order. It is simpler to take the cube root first, since $(27)^{1/3} = \sqrt[3]{27} = 3$ (because $3^3 = 27$).

$$x = (27)^{2/3} = \left[(27)^{1/3}\right]^2 = 3^2 = 9$$

Check the answer: Plug $x = 9$ into the original equation.

$$x\sqrt{x} = 27$$

$$9\sqrt{9} = 9(3) = 27$$

The answer checks out.

Problem 10

Directions: Determine what comes next in the following pattern.

$$16x^8$$

$$8x^6$$

$$4x^4$$

$$2x^2$$

❖ You can find the solution on the following page.

Solution to Problem 10

What can you do algebraically to each term in order to create the given pattern? You can divide each term by $2x^2$. For example,

$$\frac{16x^8}{2x^2} = 8x^6$$

We applied the rule that

$$\frac{x^m}{x^n} = x^{m-n}$$

Note that $\frac{x^8}{x^2} = x^{8-2} = x^6$. To determine the answer to the problem, divide the last term of the pattern by $2x^2$.

$$\frac{2x^2}{2x^2} = 1$$

The final answer is 1. Note that if your answer is $1x^0$ or x^0, your answer simplifies to 1 because $x^0 = 1$.

Check the answer: Start with 1 and multiply by $2x^2$, and see if this reproduces the same pattern in reverse.

$$(1)(2x^2) = 2x^2$$
$$(2x^2)(2x^2) = 4x^4$$
$$(4x^4)(2x^2) = 8x^6$$
$$(8x^6)(2x^2) = 16x^8$$

The answer checks out.

Problem 11

Directions: Apply algebra to solve the following problem.

Variables w, x, y, and z are related by the following equation.

$$w = 8\frac{xy}{z^2}$$

If x doubles, y triples, and z quadruples, by what factor does w change?

❖ You can find the solution on the following page.

Solution to Problem 11

We can designate initial values with a subscript 1 and final values with a subscript 2.

$$w_1 = 8\frac{x_1 y_1}{z_1^2} \quad , \quad w_2 = 8\frac{x_2 y_2}{z_2^2}$$

It is convenient to divide the second equation by the first equation. We will divide the left-hand sides and the right-hand sides.

$$\frac{w_2}{w_1} = \frac{8\frac{x_2 y_2}{z_2^2}}{8\frac{x_1 y_1}{z_1^2}} = \frac{\frac{x_2 y_2}{z_2^2}}{\frac{x_1 y_1}{z_1^2}}$$

The coefficient (8) cancels. Recall that the way to divide two fractions is to multiply the first fraction by the reciprocal of the second fraction, as in $\frac{a}{b} \div \frac{c}{d} = \frac{a\,d}{b\,c}$.

$$\frac{w_2}{w_1} = \frac{x_2 y_2}{z_2^2} \div \frac{x_1 y_1}{z_1^2} = \frac{x_2 y_2}{z_2^2}\frac{z_1^2}{x_1 y_1} = \frac{x_2 y_2 z_1^2}{x_1 y_1 z_2^2}$$

Now we will express the last sentence of the problem with algebra. The following equations represent that x doubles, y triples, and z quadruples.

$$x_2 = 2x_1 \quad , \quad y_2 = 3y_1 \quad , \quad z_2 = 4z_1$$

Plug these expressions into the previous equation.

$$\frac{w_2}{w_1} = \frac{x_2 y_2 z_1^2}{x_1 y_1 z_2^2} = \frac{(2x_1)(3y_1)z_1^2}{x_1 y_1 (4z_1)^2}$$

Note that $(4z_1)^2 = 4^2 z_1^2 = 16z_1^2$ according to the rule $(xy)^m = x^m y^m$. Also note that $x_1, y_1,$ and z_1 all cancel out. For example, $\frac{z_1^2}{z_1^2} = 1$.

$$\frac{w_2}{w_1} = \frac{6x_1 y_1 z_1^2}{16x_1 y_1 z_1^2} = \frac{3}{8}$$

Multiply both sides by w_1 to get $w_2 = \frac{3}{8}w_1$. Thus, w will be 3/8 of its initial value.

Check the answer: Try it out with numbers, such as $x_1 = 3$, $y_1 = 4$, and $z_1 = 2$.

$$x_2 = 2x_1 = 2(3) = 6 \quad , \quad y_2 = 3y_1 = 3(4) = 12 \quad , \quad z_2 = 4z_1 = 4(2) = 8$$

$$w_1 = 8\frac{x_1 y_1}{z_1^2} = 8\frac{(3)(4)}{(2)^2} = 8\left(\frac{12}{4}\right) = 24 \quad , \quad w_2 = 8\frac{x_2 y_2}{z_2^2} = 8\frac{(6)(12)}{(8)^2} = 8\left(\frac{72}{64}\right) = 9$$

Since $\frac{w_2}{w_1} = \frac{9}{24} = \frac{3}{8}$ (reduce $\frac{9}{24}$ by dividing 9 and 24 both by 3), the answer checks out.

Problem 12

Directions: Solve for x in the equation below.

$$3(x - y) + 5^{z+(2-z)} + \frac{6y - x}{2} = 30$$

❖ You can find the solution on the following page.

Solution to Problem 12

There are three variables in this equation: x, y, and z. Ordinarily, you would need three independent equations in order to solve for three unknowns. However, in this case, you can solve for x in the given equation. The only trick is not to give up before you try. Distribute the 3, simplify the exponent, and distribute the one-half. We're applying the rules $a(b-c) = ab - ac$ and $\frac{d-f}{g} = \frac{1}{g}(d-f) = \frac{d}{g} - \frac{f}{g}$.

$$3(x-y) + 5^{z+(2-z)} + \frac{6y-x}{2} = 30$$

$$3x - 3y + 5^{z+2-z} + \frac{6y}{2} - \frac{x}{2} = 30$$

Note that $z + 2 - z = 2$ (since $z - z = 0$) and $\frac{6y}{2} = 3y$ (since $\frac{6}{2} = 3$).

$$3x - 3y + 5^2 + 3y - \frac{x}{2} = 30$$

Note that $-3y + 3y = 0$. Look at that: Two of the variables (y and z) canceled out. There is just one variable left: x. Now the problem is easy.

$$3x + 5^2 - \frac{x}{2} = 10$$

Factor out the x. We're applying the rule $ax - bx = (a - b)x$. Note that $5^2 = 25$.

$$\left(3 - \frac{1}{2}\right)x + 25 = 30$$

Note that $3 - \frac{1}{2} = \frac{6}{2} - \frac{1}{2} = \frac{5}{2}$. Subtract 25 from both sides: $30 - 25 = 5$.

$$\frac{5x}{2} = 5$$

Multiply both sides of the equation by 2 and divide both sides by 5.

$$x = \frac{(5)(2)}{5} = \frac{10}{5} = 2$$

Note that y and z are both indeterminate. However, the problem only asks for x.

Check the answer: Plug $x = 2$ into the original equation (y and z cancel out again).

$$3(2-y) + 5^{z+(2-z)} + \frac{6y-2}{2} = 30$$

$$6 - 3y + 5^2 + 3y - 1 = 30$$

$$6 + 25 - 1 = 30$$

The answer checks out.

Problem 13

Directions: Solve for x in the equation below. (Don't use guess and check.)

$$x^{3/2} = \frac{32}{x^{1/6}}$$

❖ You can find the solution on the following page.

Solution to Problem 13

Multiply both sides of the equation by $x^{1/6}$.

$$x^{3/2}x^{1/6} = 32$$

Apply the rule $x^m x^n = x^{m+n}$.

$$x^{3/2+1/6} = 32$$

Add the fractions by making a common denominator.

$$\frac{3}{2} + \frac{1}{6} = \frac{3}{2}\frac{3}{3} + \frac{1}{6} = \frac{9}{6} + \frac{1}{6} = \frac{10}{6} = \frac{5}{3}$$

(Divide 10 and 6 both by 2 in order to reduce 10/6 to 5/3.) Since $3/2 + 1/6 = 5/3$, the previous equation becomes:

$$x^{5/3} = 32$$

Raise both sides of the equation to the power of 3/5. Why? We will apply the rule that $(x^m)^n = x^{mn}$ such that $(x^m)^{\frac{1}{m}} = x^{m\left(\frac{1}{m}\right)} = x^1 = x$.

$$\left(x^{\frac{5}{3}}\right)^{\frac{3}{5}} = (32)^{\frac{3}{5}}$$

Note that $\left(x^{\frac{5}{3}}\right)^{\frac{3}{5}} = x^{\frac{5}{3}\frac{3}{5}} = x^{\frac{15}{15}} = x^1 = x$.

$$x = 32^{3/5}$$

The power of 3/5 means to both raise to the 3rd power and take the 5th root, in any order. We will take the 5th root first and then cube it.

$$x = 32^{3/5} = (32^3)^{1/5} = \left(32^{1/5}\right)^3 = 2^3 = 8$$

Note that $32^{1/5} = 2$ since $2^5 = 32$. If you enter 32^(3/5) on a calculator, you can easily verify that $x = 32^{3/5} = 8$.

Check the answer: Plug $x = 8$ into the original equation. Use a calculator.

$$8^{3/2} = \frac{32}{8^{1/6}}$$

$$22.627417 \approx \frac{32}{1.414214}$$

$$22.627417 \approx 22.62741$$

The answer checks out (within reasonable rounding tolerance).

Problem 14

Directions: Apply algebra to solve the following word problem.

A rectangle has a perimeter of 15 m and an area of 9 m². Determine the length and width of the rectangle.

❖ You can find the solution on the following page.

Solution to Problem 14

Express the given information in equations. We will use x for length and y for width. The area (A) of a rectangle equals its length times its width:
$$A = xy$$
The perimeter (P) of a rectangle equals twice its length plus twice its width:
$$P = 2x + 2y$$
Plug in the given numbers for the area ($A = 9$ m^2) and perimeter ($P = 15$ m).
$$9 = xy$$
$$15 = 2x + 2y$$
Divide both sides of the first equation by x.
$$\frac{9}{x} = y$$
Substitute this expression for y into the other equation.
$$15 = 2x + 2\left(\frac{9}{x}\right) = 2x + \frac{18}{x}$$
Multiply both sides of the equation by x. Note that $2xx = 2x^2$ and $\frac{18x}{x} = 18$.
$$15x = 2x^2 + 18$$
This is a quadratic equation. First express the equation in standard form. Subtract $2x^2$ and 18 from both sides of the equation. It's convenient to then multiply by -1.
$$2x^2 - 15x + 18 = 0$$
Either factor this equation or apply the quadratic formula. We will factor it.
$$(2x - 3)(x - 6) = 0$$
There are two possible solutions:
$$2x - 3 = 0 \quad \text{or} \quad x - 6 = 0$$
The two solutions are $x = \frac{3}{2} = 1.5$ m and $x = 6$ m. Plug each solution into $\frac{9}{x} = y$. The two corresponding solutions are $y = 6$ m and $y = 1.5$ m. Either way, the rectangle has a length of 6 m and a width of 1.5 m.

Check the answers: Plug $x = 1.5$ m and $y = 6$ m into the original equations.
$$A = xy = (1.5 \text{ m})(6 \text{ m}) = 9 \text{ m}^2$$
$$P = 2x + 2y = 2(1.5 \text{ m}) + 2(6 \text{ m}) = 3 \text{ m} + 12 \text{ m} = 15 \text{ m}$$
You will get the same thing using $x = 6$ m and $y = 1.5$ m. The answers check out.

Problem 15

Directions: Apply algebra to solve the following word problem.

A girl travels with a constant speed of 4 m/s directly from point A to point B and then travels with a constant speed of 6 m/s directly from point B to point A. Given that average speed is defined as the total distance traveled by the total time, prove that the girl's average speed is 4.8 m/s. (Why is 5 m/s incorrect?)

❖ You can find the solution on the following page.

Solution to Problem 15

Since the "average value" of 4 and 6 is 5, you might expect the "average speed" to be 5 m/s, but this incorrect. Why? Because the girl spends more time traveling at the slower speed, the answer must be less than 5 m/s. As an extreme case, suppose that she traveled 4 m/s for one second and 6 m/s for three days. Would you still feel that the average should be 5 m/s? (This should convince you that time matters.)

Let's use x to represent the distance between points A and B. Then $2x$ represents the total distance traveled, since she makes a round trip. We will use t_1 and t_2 to represent the time taken for each part of the trip. The total time is $t_1 + t_2$. Average speed is defined as the total distance divided by the total time:

$$\text{ave. spd.} = \frac{\text{total dist.}}{\text{total time}} = \frac{2x}{t_1 + t_2}$$

The girl travels 4 m/s for the first trip and 6 m/s for the second trip. The distance for each trip is x. Each speed equals distance over time:

$$4 = \frac{x}{t_1} \quad , \quad 6 = \frac{x}{t_2}$$

Solve for time in each equation. Multiply both sides by time and divide by speed.

$$4t_1 = x \quad , \quad 6t_2 = x$$
$$t_1 = \frac{x}{4} \quad , \quad t_2 = \frac{x}{6}$$

Substitute these equations for time into the equation for average speed. We will add the fractions by making a common denominator (in the denominator).

$$\text{ave. spd.} = \frac{2x}{\frac{x}{4}+\frac{x}{6}} = \frac{2x}{\frac{x}{4}\frac{3}{3}+\frac{x}{6}\frac{2}{2}} = \frac{2x}{\frac{3x}{12}+\frac{2x}{12}} = \frac{2x}{5x/12}$$

To divide by a fraction, multiply by its reciprocal. The distance (x) cancels out.

$$\text{ave. spd.} = 2x \div \frac{5x}{12} = 2x\left(\frac{12}{5x}\right) = \frac{24x}{5x} = \frac{24}{5} = 4.8 \text{ m/s}$$

Check the answer: Try it with numbers. Suppose that the distance between points A and B is 96 m. The total distance traveled is then $2x = 2(96) = 192$ m. The times are $t_1 = \frac{x}{4} = \frac{96}{4} = 24$ s and $t_2 = \frac{x}{6} = \frac{96}{6} = 16$ s. The total time is $t_1 + t_2 = 24 + 16 = 40$ s. The average speed is $\frac{192}{40} = 4.8$ m/s. The answer checks out.

Problem 16

Directions: The equation given below applies to parts (A) and (B) of this problem. In the equation below, x, y, and z are all positive and $x \neq y$.

$$\frac{x - z}{y - x} = \frac{x + z}{y + x}$$

(A) Apply algebra to derive the equation below from the equation given above.

$$\frac{x - z}{y - x} = \frac{z}{x}$$

(B) Also derive the equation below.

$$\frac{x - z}{y - x} = \frac{x}{y}$$

❖ You can find the solution on the following page.

Solution to Problem 16

(A) Cross multiply. Recall that $\frac{a}{b} = \frac{c}{d}$ becomes $ad = bc$ when you cross multiply.
$$(x - z)(y + x) = (y - x)(x + z)$$
Apply the f.o.i.l. method.
$$xy + x^2 - yz - xz = xy + yz - x^2 - xz$$
Combine like terms. Note that the xy and xz terms cancel out.
$$2x^2 = 2yz$$
Divide both sides of this equation by $2z$.
$$\frac{x^2}{z} = y$$
Plug this expression in for y on the right-hand side of the given equation.
$$\frac{x - z}{y - x} = \frac{x + z}{\frac{x^2}{z} + x}$$
Multiply the right-hand side by $\frac{z}{z}$ (which equates to multiplying by one). Distribute the z in the denominator. In the last step, we factor x out of the denominator.
$$\frac{x - z}{y - x} = \frac{(x + z)z}{\left(\frac{x^2}{z} + x\right)z} = \frac{(x + z)z}{\frac{x^2 z}{z} + xz} = \frac{(x + z)z}{x^2 + xz} = \frac{(x + z)z}{(x + z)x} = \frac{z}{x}$$
(Since the problem states that x, y, and z are all positive and also states that $x \neq y$, we don't need to worry about dividing by zero.)

(B) Recall the equation $2x^2 = 2yz$. Divide both sides of this equation by $2xy$.
$$\frac{x}{y} = \frac{z}{x}$$
Substitute this into the equation that we derived in part (A).
$$\frac{x - z}{y - x} = \frac{z}{x} = \frac{x}{y}$$

Check the answers: Try it with numbers. Suppose that $x = 6$ and $y = 2$. According to the equation $\frac{x}{y} = \frac{z}{x}$ that we found, $\frac{6}{2} = 3 = \frac{z}{6}$ such that $z = (3)(6) = 18$.
$$\frac{x - z}{y - x} = \frac{6 - 18}{2 - 6} = \frac{-12}{-4} = 3 \quad \text{and} \quad \frac{x + z}{y + x} = \frac{6 + 18}{2 + 6} = \frac{24}{8} = 3$$
The solution checks out.

Problem 17

Directions: Solve for all possible nonzero values of x in the equation below.

$$4x^3 - 4x^2\sqrt{3} = 24x$$

❖ You can find the solution on the following page.

Solution to Problem 17

Subtract $24x$ from both sides of the equation and divide by 4.

$$x^3 - x^2\sqrt{3} - 6x = 0$$

Factor out an x.

$$x(x^2 - x\sqrt{3} - 6) = 0$$

Either $x = 0$ or $x^2 - x\sqrt{3} - 6 = 0$. Since the problem specifically states that we are looking for nonzero values of x, only the second case applies.

$$x^2 - x\sqrt{3} - 6 = 0$$

This is a quadratic equation. Compare this to the general equation $ax^2 + bx + c = 0$ to determine that:

$$a = 1 \quad , \quad b = -\sqrt{3} \quad , \quad c = -6$$

Plug these values into the quadratic formula. Note that $b^2 = (-\sqrt{3})^2 = +3$ and that $-b = -(-\sqrt{3}) = +\sqrt{3}$.

$$x = \frac{-b \pm \sqrt{b^2 - 4ac}}{2a} = \frac{-(-\sqrt{3}) \pm \sqrt{(-\sqrt{3})^2 - 4(1)(-6)}}{2(1)} = \frac{\sqrt{3} \pm \sqrt{3 + 24}}{2}$$

Note that the two minus signs make a plus sign: $-4(1)(-6) = (-4)(-6) = +24$. Also note that $\sqrt{27} = \sqrt{(3)(9)} = \sqrt{3}\sqrt{9} = 3\sqrt{3}$ (we factored out a perfect square).

$$x = \frac{\sqrt{3} \pm \sqrt{27}}{2} = \frac{\sqrt{3} \pm 3\sqrt{3}}{2}$$

There are two possible nonzero answers for x:

$$x = \frac{\sqrt{3} + 3\sqrt{3}}{2} = \frac{4\sqrt{3}}{2} = 2\sqrt{3} \quad \text{or} \quad x = \frac{\sqrt{3} - 3\sqrt{3}}{2} = \frac{-2\sqrt{3}}{2} = -\sqrt{3}$$

Check the answers: Plug $x = 2\sqrt{3}$ and $x = -\sqrt{3}$ into the original equation.

$$4x^3 - 4x^2\sqrt{3} = 24x$$

$$4(2\sqrt{3})^3 - 4(2\sqrt{3})^2\sqrt{3} = 24(2\sqrt{3}) \quad \text{or} \quad 4(-\sqrt{3})^3 - 4(-\sqrt{3})^2\sqrt{3} = 24(-\sqrt{3})$$

$$4(8)(3\sqrt{3}) - 4(4)(3)\sqrt{3} = 48\sqrt{3} \quad \text{or} \quad 4(-3\sqrt{3}) - 4(3)\sqrt{3} = -24\sqrt{3}$$

$$96\sqrt{3} - 48\sqrt{3} = 48\sqrt{3} \quad \text{or} \quad -12\sqrt{3} - 12\sqrt{3} = -24\sqrt{3}$$

$$48\sqrt{3} = 48\sqrt{3} \quad \text{or} \quad -24\sqrt{3} = -24\sqrt{3}$$

The answers check out.

Problem 18

Directions: Apply algebra to solve the following word problem. Don't make up a number for the amount of the investment. (Assume that there are no other fees, taxes, or benefits involved other than what is explicitly stated.)

A businessman invests a certain amount of money in a business venture and loses 20% of his money. He invests the remaining amount of money in another business venture and that money increases by 20%. Determine the ratio of the amount of money that he has now compared to his initial investment. (Why is 1:1 incorrect?)

❖ You can find the solution on the following page.

Solution to Problem 18

Since the first investment decreases by 20% and the second investment increases by 20%, you might wonder why they don't effectively cancel out and make the ratio one to one (1:1). The reason is that the first 20% applies to a larger amount and the second 20% applies to a smaller amount. (We'll see this when we check the answer with numbers later. First, as directed in the problem, we need to solve the problem without making up a number for the amount of the investment.)

We define the following variables:
- x is the amount of the initial investment.
- y is the amount of money remaining after the first investment.
- z is the amount of money remaining after both investments have been made.

We wish to find the ratio of z to x (which we can express in the form $\frac{z}{x}$ or $z:x$).

The first investment results in a 20% reduction. Divide by 100% in order to convert 20% from a percent to a decimal: $20\% = 20\%/100\% = 0.2$.
$$y = (1 - 20\%)x = (1 - 0.2)x = 0.8x$$
We see that a 20% reduction equates to multiplying by 80%.

The second investment results in a 20% increase.
$$z = (1 + 20\%)y = (1 + 0.2)y = 1.2y$$
Note that the first 20% reduces x, whereas the second 20% increases y. This is what we meant earlier about the 20%'s applying to different amounts. Plug the equation $y = 0.8x$ (which we found earlier) into the previous equation.
$$z = 1.2y = 1.2(0.8x) = 0.96x$$
Divide both sides of the equation by x to see that $\frac{z}{x} = 0.96$. The ratio is 0.96, which we could alternatively express as 96:100 or 24:25. (Overall, this is a loss of 4%.)

Check the answer: Try it with numbers. Assume the initial investment is $x = \$100$. When the first investment loses 20%, what remains is $y = \$80$. When the second investment increases by 20%, we must raise $80 by 20%. Note that 20% of $80 is $16 (since $\$80 \times 0.2 = \$16$), such that $z = \$96$ (add $16 to $80 to get this). The final amount ($96) is 0.96 times the initial investment ($100). The answer checks out.

Problem 19

Directions: Solve for x and y in the system of equations below.

$$4\sqrt{x} + 3\sqrt{y} = 43$$

$$5\sqrt{x} - 2\sqrt{y} = 25$$

❖ You can find the solution on the following page.

Solution to Problem 19

If we define $t = \sqrt{x}$ and $u = \sqrt{y}$, the given equations look like this:

$$4t + 3u = 43$$
$$5t - 2u = 25$$

Without the squareroots, the problem seems simpler. One way to solve this system is to setup simultaneous equations. We will multiply the top equation by 2 and the bottom equation by 3.

$$2(4t + 3u) = 2(43) \quad \rightarrow \quad 8t + 6u = 86$$
$$3(5t - 2u) = 3(25) \quad \rightarrow \quad 15t - 6u = 75$$

Now if we add the two equations together, the $6u$'s will cancel out.

$$8t + 6u + 15t - 6u = 86 + 75$$
$$23t = 161$$
$$t = \frac{161}{23} = 7$$

Plug the answer for t into one of the equations at the top of the page.

$$4(7) + 3u = 43$$
$$28 + 3u = 43$$
$$3u = 43 - 28 = 15$$
$$u = \frac{15}{3} = 5$$

We're not finished yet because we need to solve for x and y. Square the equations $t = \sqrt{x}$ and $u = \sqrt{y}$ to obtain $t^2 = x$ and $u^2 = y$.

$$x = t^2 = 7^2 = 49 \quad \text{and} \quad y = u^2 = 5^2 = 25$$

Check the answers: Plug $x = 49$ and $y = 25$ into the original equations.

$$4\sqrt{x} + 3\sqrt{y} = 4\sqrt{49} + 3\sqrt{25} = 4(7) + 3(5) = 28 + 15 = 43$$
$$5\sqrt{x} - 2\sqrt{y} = 5\sqrt{49} - 2\sqrt{25} = 5(7) - 2(5) = 35 - 10 = 25$$

The answers check out.

Problem 20

Directions: Find two different values for x which satisfy the following equation.

$$\frac{x}{8} - \frac{2}{x} = 0$$

❖ You can find the solution on the following page.

Solution to Problem 20

Add $\frac{2}{x}$ to both sides of the equation.

$$\frac{x}{8} = \frac{2}{x}$$

Cross multiply. Recall that $\frac{a}{b} = \frac{c}{d}$ becomes $ad = bc$ when you cross multiply. Also note that $xx = x^2$.

$$x^2 = (8)(2)$$
$$x^2 = 16$$

Squareroot both sides of the equation.

$$x = \pm\sqrt{16}$$
$$x = \pm 4$$

The two possible answers are $x = -4$ and $x = 4$. Note that $x = -4$ works because $(-4)^2 = +16$ (the minus sign gets squared).

Check the answers: Plug $x = -4$ and $x = 4$ into the original equation.

$$\frac{x}{8} - \frac{2}{x} = 0$$

$$-\frac{4}{8} - \frac{2}{-4} = 0 \quad \text{or} \quad \frac{4}{8} - \frac{2}{4} = 0$$

$$-\frac{1}{2} + \frac{1}{2} = 0 \quad \text{or} \quad \frac{1}{2} - \frac{1}{2} = 0$$

The answers check out. Note that $-\frac{2}{-4} = -\left(-\frac{1}{2}\right) = +\frac{1}{2}$.

Problem 21

Directions: Explain precisely what is wrong with the following 'proof.'

$$(-1)(-1) = (1)(1)$$

$$\sqrt{(-1)(-1)} = \sqrt{(1)(1)}$$

$$-1 = 1$$

❖ You can find the explanation on the following page.

Solution to Problem 21

The last line, $-1 = 1$, is obviously incorrect. The problem is to figure out why the attempted proof resulted in that incorrect conclusion.

The first line, $(-1)(-1) = (1)(1)$, is certainly correct since $(-1)(-1) = 1$ and since $(1)(1) = 1$. The middle line, $\sqrt{(-1)(-1)} = \sqrt{(1)(1)}$, is correct for the same reason that the first line is correct. Since $(-1)(-1) = 1$ and $(1)(1) = 1$, the middle line is equivalent to writing $\sqrt{1} = \sqrt{1}$. Therefore, there must be a mistake going from the middle line to the last line, but what exactly is the mistake?

Do you think the mistake is going from $\sqrt{(-1)(-1)}$ to -1 on the left-hand side? Be careful here. This equates to saying that $\sqrt{xx} = \sqrt{x^2} = x$, where $x = -1$. Do you see a mistake with that? Let's approach this another way: $\sqrt{(-1)(-1)} = \sqrt{1}$. This means, "what number times itself equals 1?" It's not incorrect to state that the squareroot of one equals minus one: $\sqrt{1} = -1$. Why? Because it's true that $(-1)(-1) = 1$.

Really, 'incorrect' is the wrong word. The right word is 'incomplete.' It isn't incorrect to write $\sqrt{(-1)(-1)} = -1$, but it is incomplete. The reason that it's incomplete is that there are two answers to $\sqrt{(-1)(-1)}$. It's both correct and complete to write $\sqrt{(-1)(-1)} = \pm 1$. But that's not the only problem! It's equally incomplete to write $\sqrt{(1)(1)} = 1$. We should similarly write $\sqrt{(1)(1)} = \pm 1$. In fact, both equations are the same. Since $(-1)(-1) = (1)(1) = 1$, in both cases we could write $\sqrt{1} = \pm 1$. In general, we should write $\sqrt{x^2} = \pm x$.

That's the real issue here. The last line of the proof should be $\pm 1 = \pm 1$. (Some math classes adopt the convention of ignoring the negative roots, but in some physical situations that negative root has significance, like the sign of the final velocity. For example, if you throw a rock upward and call the upward direction positive, when we solve for the final velocity with the formula $v = \sqrt{v_0^2 + 2a\Delta x}$, a positive root indicates that the rock is still traveling upward, whereas a negative root indicates that the rock is now falling downward.)

Problem 22

Directions: Solve for x in the equation below, where x^2 and $\frac{3x\sqrt{3}}{2}$ are both exponents.

$$2^{x^2} 2^{\frac{3x\sqrt{3}}{2}} = 8$$

❖ You can find the solution on the following page.

Solution to Problem 22

The trick is to realize that $2^3 = 8$. This allows us to rewrite the given equation as:

$$2^{x^2} 2^{\frac{3x\sqrt{3}}{2}} = 2^3$$

This has the same structure as $2^p 2^q = 2^r$ where $p = x^2$, $q = \frac{3x\sqrt{3}}{2}$, and $r = 3$. Apply the rule that $x^m x^n = x^{m+n}$ to get $2^{p+q} = 2^r$. This can only be true if $p + q = r$. Plug the expressions $p = x^2$, $q = \frac{3x\sqrt{3}}{2}$, and $r = 3$ into the equation $p + q = r$.

$$x^2 + \frac{3x\sqrt{3}}{2} = 3$$

This is a quadratic equation. Subtract 3 from both sides and multiply both sides by 2.

$$2x^2 + 3x\sqrt{3} - 6 = 0$$

Compare this to the general equation $ax^2 + bx + c = 0$ to determine that $a = 2$, $b = 3\sqrt{3}$, and $c = -6$. Plug these values into the quadratic formula. Note that $b^2 = \left(3\sqrt{3}\right)^2 = (3)^2\left(\sqrt{3}\right)^2 = (9)(3) = 27$.

$$x = \frac{-b \pm \sqrt{b^2 - 4ac}}{2a} = \frac{-3\sqrt{3} \pm \sqrt{\left(3\sqrt{3}\right)^2 - 4(2)(-6)}}{2(2)} = \frac{-3\sqrt{3} \pm \sqrt{27 + 48}}{4}$$

Note that the two minus signs make a plus sign: $-4(2)(-6) = (-8)(-6) = 48$. Also note that $\sqrt{75} = \sqrt{(3)(25)} = \sqrt{3}\sqrt{25} = 5\sqrt{3}$ (we factored out a perfect square).

$$x = \frac{-3\sqrt{3} \pm \sqrt{75}}{4} = \frac{-3\sqrt{3} \pm 5\sqrt{3}}{4}$$

$$x = \frac{-3\sqrt{3} - 5\sqrt{3}}{4} = \frac{-8\sqrt{3}}{4} = -2\sqrt{3} \quad \text{or} \quad x = \frac{-3\sqrt{3} + 5\sqrt{3}}{4} = \frac{2\sqrt{3}}{4} = \frac{\sqrt{3}}{2}$$

Check the answers: Plug $x = -2\sqrt{3}$ and $x = \frac{\sqrt{3}}{2}$ into the original equation.

$$2^{\left(-2\sqrt{3}\right)^2} 2^{3\left(-2\sqrt{3}\right)\sqrt{3}/2} = 2^3 \quad \text{or} \quad 2^{\left(\sqrt{3}/2\right)^2} 2^{3\left(\sqrt{3}/2\right)\sqrt{3}/2} = 2^3$$

$$2^{12} 2^{-9} = 2^3 \quad \text{or} \quad 2^{3/4} 2^{9/4} = 2^{12/4} = 2^3$$

The answers check out.

Problem 23

Directions: Solve for two different values for x that satisfy the following equation.

$$\frac{16}{x^2} = \frac{25}{(x+18)^2}$$

❖ You can find the solution on the following page.

Solution to Problem 23

Cross multiply. Recall that $\frac{a}{b} = \frac{c}{d}$ becomes $ad = bc$ when you cross multiply.

$$16(x + 18)^2 = 25x^2$$

There are two ways to approach the above equation. If you write $(x + 18)^2 = x^2 + 36x + 324$, you will need to solve a quadratic equation. In this case, there is a way to avoid the quadratic. Squareroot both sides of the equation instead.

$$\sqrt{16(x + 18)^2} = \sqrt{25x^2}$$

However, if you do this, you won't get two answers for x (like you would using the quadratic) unless you remember to include $\pm$ signs with one of your squareroots.

$$\sqrt{16(x + 18)^2} = \pm\sqrt{25x^2}$$

Apply the rule $\sqrt{ab} = \sqrt{a}\sqrt{b}$, which is the same as $(ab)^m = a^m b^m$ with $m = 1/2$.

$$\sqrt{16}\sqrt{(x + 18)^2} = \pm\sqrt{25}\sqrt{x^2}$$

Note that $\sqrt{x^2} = x$ and $\sqrt{(x + 18)^2} = x + 18$ (we already included a $\pm$ earlier).

$$4(x + 18) = \pm 5x$$
$$4x + 72 = \pm 5x$$

$$4x + 72 = 5x \quad \text{or} \quad 4x + 72 = -5x$$
$$72 = 5x - 4x \quad \text{or} \quad 72 = -5x - 4x$$
$$72 = x \quad \text{or} \quad 72 = -9x$$
$$72 = x \quad \text{or} \quad -8 = x$$

Check the answers: Plug $x = -8$ and $x = 72$ into the original equation.

$$\frac{16}{x^2} = \frac{25}{(x + 18)^2}$$

$$\frac{16}{(-8)^2} = \frac{25}{(-8 + 18)^2} \quad \text{or} \quad \frac{16}{(72)^2} = \frac{25}{(72 + 18)^2}$$

$$\frac{16}{64} = \frac{25}{(10)^2} \quad \text{or} \quad \frac{16}{5184} = \frac{25}{(90)^2}$$

$$\frac{16}{64} = \frac{25}{100} \quad \text{or} \quad \frac{16}{5184} = \frac{25}{8100}$$

$$0.25 = 0.25 \quad \text{or} \quad 0.00308642 = 0.00308642$$

The answers check out (using a calculator for the last step).

Problem 24

Directions: Represent the following word problem with an algebraic equation.

A philanthropist has one million dollars. Every day, the philanthropist gives away one-half of his money. Write an equation representing the number of days that will pass until the philanthropist has just one dollar left. (Assume that there are no other sources of income, expenses, or taxes involved.)

❖ You can find the solution on the following page.

Solution to Problem 24

The philanthropist initially has $1,000,000. According to the instructions, you must apply algebra to represent the word problem with an equation. That is, you're not allowed to keep dividing $1,000,000 by 2 until you reach $1. (However, we will do that later to check our answer.) How can you divide by 2 an unknown number of times? Let x represent the number of days and let y represent the money that the philanhtropist has left over. The solution is to divide $1,000,000 by 2^x.

$$y = \frac{\$1,000,000}{2^x} \quad \text{or} \quad y = \$1,000,000\left(\frac{1}{2}\right)^x$$

That completely answers the question. Note that the directions didn't ask us to solve for x. However, if you know about logarithms, you can solve the equation as follows:

$$x = \log_2 1,000,000 = \frac{\log_{10} 1,000,000}{\log_{10} 2} \approx 19.93 \approx 20$$

The answer is 20 days. We used the change of base formula, $\log_a x = \frac{\log_b x}{\log_b a}$.

Check the answer: Divide $1,000,000 by 2 repeatedly until you reach $1.

Day	Money	Day	Money
1	$500,000.00	11	$488.28
2	$250,000.00	12	$244.14
3	$125,000.00	13	$122.07
4	$62,500.00	14	$61.04
5	$31,250.00	15	$30.52
6	$15,625.00	16	$15.26
7	$7,812.50	17	$7.63
8	$3,906.25	18	$3.81
9	$1,953.13	19	$1.91
10	$976.56	20	$0.95

Compare the table above with the formula, using $x = 20$ with a calculator.

$$y = \frac{\$1,000,000}{2^{20}} = \frac{\$1,000,000}{1,048,576} = \$0.95$$

The answer checks out since $0.95 is approximately equal to $1.00.

Problem 25

Directions: Derive the indicated inequality from the relationships given below.

$$\frac{1}{x} + \frac{1}{y} = \frac{1}{6}$$

$$z = y + x \quad , \quad w = y - x$$

$$x > 0 \quad , \quad y > x$$

Use the equations and inequalities above to derive the following inequality:

$$z > 24$$

❖ You can find the solution on the following page.

Solution to Problem 25

Solve for x and y in terms of w and z. Add the equations for w and z to get $z + w = y + x + y - x = 2y$ and subtract w from z to get $z - w = y + x - (y - x) = y + x - y + x = 2x$. Note that $-(y - x) = -y - (-x) = -y + x$. Divide each equation by 2.

$$y = \frac{z + w}{2} \quad , \quad x = \frac{z - w}{2}$$

Substitute these expressions into the equation $\frac{1}{x} + \frac{1}{y} = \frac{1}{6}$.

$$\frac{2}{z - w} + \frac{2}{z + w} = \frac{1}{6}$$

Make a common denominator. Note that $(a - b)(a + b) = a^2 - b^2$.

$$\frac{2}{z - w}\left(\frac{z + w}{z + w}\right) + \frac{2}{z + w}\left(\frac{z - w}{z - w}\right) = \frac{1}{6}$$

$$\frac{2(z + w + z - w)}{z^2 - w^2} = \frac{1}{6}$$

$$\frac{4z}{z^2 - w^2} = \frac{1}{6}$$

Cross multiply. Recall that $\frac{a}{b} = \frac{c}{d}$ becomes $ad = bc$ when you cross multiply.

$$24z = z^2 - w^2$$

Note that z^2 is greater than $z^2 - w^2$ (since w^2 is inherently positive).

$$z^2 > z^2 - w^2 = 24z$$

$$z^2 > 24z$$

$$z > 24$$

This problem has a practical application: A convex lens can only yield a real image if the distance between the object and screen is at least 4 times the focal length. (Here the focal length is 6 and the 24 in the inequality represents 4 times the focal length.)

Check the answer: Note that $\frac{1}{y} = \frac{1}{6} - \frac{1}{x} = \frac{x - 6}{6x}$ and $y = \frac{6x}{x - 6}$. Let's try some numbers.

- If $x = 8$, then $y = \frac{(6)(8)}{8 - 6} = 24$. Here, $z = x + y = 32$, which is greater than 24.
- If $x = 9$, then $y = \frac{(6)(9)}{9 - 6} = 18$. Here, $z = x + y = 27$, which is greater than 24.
- If $x = 12$, then $y = \frac{(6)(12)}{12 - 6} = 12$. Here, $z = x + y = 24$ is exactly 24, but this limiting case isn't allowed because one of the requirements was $y > x$.

Problem 26

Directions: Factor the 4 out of the following expression.

$$(4x^2 + 9)^{-3/2}$$

Your final answer should have the following form (where you fill in the blanks).

$$\underline{\hspace{1cm}}(x^2 + \underline{\hspace{1cm}})^{-3/2}$$

❖ You can find the solution on the following page.

Solution to Problem 26

Begin by factoring the 4 out of the expression $4x^2 + 9$.

$$4x^2 + 9 = 4\left(x^2 + \frac{9}{4}\right)$$

If you distribute the 4 on the right-hand side, you should see that it works. We're not finished yet, since we still need to deal with the power of $-3/2$. Apply the rule that $(ay)^m = a^m y^m$, with $a = 4$, $y = x^2 + \frac{9}{4}$, and $m = -3/2$.

$$(4x^2 + 9)^{-3/2} = \left[4\left(x^2 + \frac{9}{4}\right)\right]^{-3/2} = (4)^{-3/2}\left(x^2 + \frac{9}{4}\right)^{-3/2}$$

Note that $(4)^{-3/2} = \left(4^{1/2}\right)^{-3}$ according to the rule $(a^p)^q = a^{pq}$. Note that $4^{1/2} = 2$ such that $(4)^{-3/2} = \left(4^{1/2}\right)^{-3} = 2^{-3} = \frac{1}{2^3} = \frac{1}{8}$. Alternatively, if you enter 4^(−3/2) on a calculator, you should find that it equals 1/8. The above expression becomes:

$$\frac{1}{8}\left(x^2 + \frac{9}{4}\right)^{-3/2}$$

This is what the problem asked for. This expression has the same form as

$$\underline{\quad}(x^2 + \underline{\quad})^{-3/2}$$

where the first blank equals 1/8 and the second blank equals 9/4.

Check the answer: Try using numbers to see that this works. For example, let $x = 3$. Use a calculator.

$$[4(3)^2 + 9]^{-3/2} = [4(9) + 9]^{-3/2} = (36 + 9)^{-3/2} = 45^{-3/2} \approx 0.00331269$$

$$\frac{1}{8}\left[(3)^2 + \frac{9}{4}\right]^{-3/2} = \frac{1}{8}\left(9 + \frac{9}{4}\right)^{-3/2} = \frac{1}{8}(9 + 2.25)^{-3/2} = \frac{1}{8}(11.25)^{-3/2} \approx 0.00331269$$

The answer checks out.

Problem 27

Directions: Solve for x in the system of equations below.

$$2y - 3x - z = 20$$

$$\frac{z}{2} + 4x - y = 5$$

❖ You can find the solution on the following page.

Solution to Problem 27

There are three variables in these equations: x, y, and z. Ordinarily, you would need three independent equations in order to solve for three unknowns. However, in this case, you can solve for x. The only trick is not to give up before you try.

Multiply both sides of the second equation by 2.
$$z + 8x - 2y = 10$$
Add this equation to the first equation.
$$2y - 3x - z + z + 8x - 2y = 20 + 10$$
Note that y and z both cancel out. All that remains is:
$$5x = 30$$
Divide both sides of this equation by 5.
$$x = \frac{30}{5} = 6$$
Note that y and z are both indeterminate. However, the problem only asks for x.

Check the answer: Plug $x = 6$ into the original equations.
$$2y - 3(6) - z = 20 \quad \text{and} \quad \frac{z}{2} + 4(6) - y = 5$$
$$2y - 18 - z = 20 \quad \text{and} \quad \frac{z}{2} + 24 - y = 5$$
$$2y - z = 20 + 18 \quad \text{and} \quad \frac{z}{2} - y = 5 - 24$$
$$2y - z = 38 \quad \text{and} \quad \frac{z}{2} - y = -19$$
Multiply both sides of the second equation by -2.
$$2y - z = 38 \quad \text{and} \quad -z + 2y = 38$$
Note that $2y - z = -z + 2y$. In both cases, $2y - z$ equals 38. The answer checks out.

Problem 28

Directions: Solve for x in the equation below. (Don't use guess and check.)

$$\frac{54}{x} = x - 3$$

❖You can find the solution on the following page.

Solution to Problem 28

Multiply both sides of the given equation by x. Note that $\frac{x}{x} = 1$ and $xx = x^2$.

$$54 = x^2 - 3x$$

This is a quadratic equation. First express the equation in standard form. Subtract x^2 from both sides, add $3x$ to both sides, and then multiply by -1. (Alternatively, subtract 54 from both sides, but then to get the equation below, swap the left and right sides of the equation.)

$$x^2 - 3x - 54 = 0$$

Either factor this equation or apply the quadratic formula. We can factor it, noting that $-54 = (-9)(6)$ and that $-9x + 6x = -3x$.

$$(x - 9)(x + 6) = 0$$

The two solutions are $x = 9$ and $x = -6$. If instead you apply the quadratic formula, you will get the same answers. Compare the equation $x^2 - 3x - 54 = 0$ to the general form $ax^2 + bx + c = 0$ to see that $a = 1$, $b = -3$, and $c = -54$.

$$x = \frac{-(-3) \pm \sqrt{(-3)^2 - 4(1)(-54)}}{2(1)} = \frac{3 \pm \sqrt{9 + 216}}{2} = \frac{3 \pm \sqrt{225}}{2} = \frac{3 \pm 15}{2}$$

$$x = \frac{3 + 15}{2} = \frac{18}{2} = 9 \quad \text{or} \quad x = \frac{3 - 15}{2} = -\frac{12}{2} = -6$$

Check the answers: Plug $x = -6$ and $x = 9$ into the original equation.

$$\frac{54}{-6} = -6 - 3 \quad \text{or} \quad \frac{54}{9} = 9 - 3$$

$$-9 = -9 \quad \text{or} \quad 6 = 6$$

The answers check out.

Problem 29

Directions: Solve for x in the equation below. (Don't use guess and check.)

$$3^{x^2} = 81$$

❖ You can find the solution on the following page.

Solution to Problem 29

(Even though this problem specifically directs you not to use guess and check, most teachers don't give full credit for guess and check solutions when known techniques could be used to solve for an answer, regardless of the wording.)

Note that it's incorrect to squareroot both sides. Squarerooting both sides equates to raising both sides to the power of 1/2. It would yield $\left(3^{x^2}\right)^{1/2} = (81)^{1/2}$, simplifying to $3^{x^2/2} = (81)^{1/2}$ according to the rule $(a^b)^c = a^{bc}$. The left-hand side wouldn't reduce to 3^x.

The trick is to realize that 81 is a power of 3. Specifically, $3^4 = 81$.

$$3^{x^2} = 3^4$$

Now the equation looks like $a^y = a^b$, where $y = x^2$ and $b = 4$. The only way the above equation can be true is if:

$$x^2 = 4$$

Squareroot both sides of the equation.

$$x = \pm 2$$

Why $\pm$? That's because $(2)^2 = 4$ and $(-2)^2 = 4$.

Check the answers: Plug $x = \pm 2$ into the original equation.

$$3^{x^2} = 81$$
$$3^{(\pm 2)^2} = 81$$
$$3^4 = 81$$

The answers check out.

Problem 30

Directions: Solve for x in the equation below.

$$\frac{\dfrac{8}{x^9} \div \dfrac{x^2}{2}}{\dfrac{9}{x^3} \div \dfrac{x}{6}} \div \frac{\dfrac{x^6}{3} \div \dfrac{x^4}{2}}{\dfrac{x^5}{4} \div \dfrac{x}{6}} = \frac{24}{x^7}$$

❖ You can find the solution on the following page.

Solution to Problem 30

The way to divide two fractions is to multiply the first fraction by the reciprocal of the second fraction, as in $\frac{a}{b} \div \frac{c}{d} = \frac{a}{b}\frac{d}{c}$.

$$\frac{\dfrac{8}{x^9}\dfrac{2}{x^2}}{\dfrac{9}{x^3}\dfrac{6}{x}} \div \frac{\dfrac{x^6}{3}\dfrac{2}{x^4}}{\dfrac{x^5}{4}\dfrac{6}{x}} = \frac{24}{x^7}$$

Apply the rules $x^m x^n = x^{mn}$ and $\dfrac{x^m}{x^n} = x^{m-n}$.

$$\frac{\dfrac{16}{x^{11}}}{\dfrac{54}{x^4}} \div \frac{\dfrac{2x^2}{3}}{\dfrac{6x^4}{4}} = \frac{24}{x^7}$$

$$\left(\frac{16}{x^{11}} \div \frac{54}{x^4}\right) \div \left(\frac{2x^2}{3} \div \frac{6x^4}{4}\right) = \frac{24}{x^7}$$

$$\left(\frac{16}{x^{11}}\frac{x^4}{54}\right) \div \left(\frac{2x^2}{3}\frac{4}{6x^4}\right) = \frac{24}{x^7}$$

$$\left(\frac{8}{27x^7}\right) \div \left(\frac{8}{18x^2}\right) = \frac{24}{x^7}$$

$$\frac{8}{27x^7}\frac{18x^2}{8} = \frac{24}{x^7}$$

$$\frac{2}{3x^5} = \frac{24}{x^7}$$

$$x^2 = 36$$

$$x = \pm 6$$

Check the answers: Plug $x = \pm 6$ into the original equation. Use a calculator.

$$\frac{\dfrac{8}{6^9}\div\dfrac{6^2}{2}}{\dfrac{9}{6^3}\div\dfrac{6}{6}} \div \frac{\dfrac{6^6}{3}\div\dfrac{6^4}{2}}{\dfrac{6^5}{4}\div\dfrac{6}{6}} = \frac{24}{6^7} \quad \rightarrow \quad \frac{0.000000794 \div 18}{0.041666667 \div 1} \div \frac{15{,}552 \div 648}{1944 \div 1} \approx 0.000085734$$

$$\frac{0.000000044}{0.041666667} \div \frac{24}{1944} \approx 0.000085734$$

$$0.000001059 \div 0.012345679 \approx 0.000085734$$

$$0.000085752 \approx 0.000085734$$

The answers check out (apart from a little round-off error).

Problem 31

Directions: Find nonzero solutions for x and y in the system of equations below.

$$9x - 6y = 2x^2$$

$$\frac{3}{x} = \frac{2}{y}$$

❖ You can find the solution on the following page.

Solution to Problem 31

(Although $x = y = 0$ would solve the problem, the instructions specifically ask for nonzero solutions.) Cross multiply. Recall that $\frac{a}{b} = \frac{c}{d}$ becomes $ad = bc$ when you cross multiply.

$$3y = 2x$$
$$y = \frac{2x}{3}$$

Plug this expression into the other equation.

$$9x - 6\left(\frac{2x}{3}\right) = 2x^2$$
$$9x - 4x = 2x^2$$

Combine like terms.

$$5x = 2x^2$$

Bring all of the terms to the same side.

$$0 = 2x^2 - 5x$$

Factor out an x.

$$0 = x(2x - 5)$$

The two solutions for x are $x = 0$ and $2x - 5 = 0$. The problem specifically asked for the nonzero solution.

$$2x = 5$$
$$x = \frac{5}{2}$$

Plug this answer into the equation for y near the top of this page.

$$y = \frac{2x}{3} = \frac{2}{3}\left(\frac{5}{2}\right) = \frac{5}{3}$$

Check the answers: Plug $x = \frac{5}{2} = 2.5$ and $y = \frac{5}{3} \approx 1.66667$ into the original equations.

$$9(2.5) - 6(1.66667) = 2(2.5)^2 \quad \text{and} \quad \frac{3}{2.5} = \frac{2}{1.66667}$$
$$22.5 - 10.00002 \approx 12.5 \quad \text{and} \quad 1.2 \approx 1.1999976$$

Use a calculator. The answers check out (apart from a little round-off error).

Problem 32

Directions: Apply algebra to solve the following word problem.

A positive real number is one greater than its reciprocal. What is the number?

❖ You can find the solution on the following page.

Solution to Problem 32

Use x to represent the unknown number. Its reciprocal is $\frac{1}{x}$. The problem states that x is 1 greater than $\frac{1}{x}$.

$$x = 1 + \frac{1}{x}$$

Multiply both sides of the equation by x. Note that $xx = x^2$ and $\frac{x}{x} = 1$.

$$x^2 = x + 1$$

Bring all of the terms to the same side of the equation.

$$x^2 - x - 1 = 0$$

This is a quadratic equation. Compare the equation above to the general form $ax^2 + bx + c = 0$ to see that $a = 1$, $b = -1$, and $c = -1$.

$$x = \frac{-(-1) \pm \sqrt{(-1)^2 - 4(1)(-1)}}{2(1)} = \frac{1 \pm \sqrt{1 + 4}}{2} = \frac{1 \pm \sqrt{5}}{2}$$

The problem states that x is positive. Only the $+$ sign yields a positive answer.

$$x = \frac{1 + \sqrt{5}}{2} \approx 1.618033989$$

The decimal form was determined using a calculator.

Check the answer: Reread the problem and see if our answer satisfies the problem.

- Our answer for the positive real number is $x \approx 1.618033989$.
- The reciprocal of this positive real number is $\frac{1}{x} \approx 0.618033989$.
- Our answer is one greater than its reciprocal:

$$x - \frac{1}{x} \approx 1.618033989 - 0.618033989 = 1$$

The answer checks out.

Problem 33

Directions: Solve for x in the equation below. (Don't use guess and check.)

$$8 - [2 - (4 - x)] = -4[x - 3(9 - x)] - 8$$

❖ You can find the solution on the following page.

Solution to Problem 33

The main idea behind the solution to this problem is to be careful distributing the minus signs. Recall the rules $-(x + y) = -x - y$ and $-(x - y) = -x + y$. Note that the 8's do not cancel (since they have different signs).

$$8 - [2 - (4 - x)] = -4[x - 3(9 - x)] - 8$$
$$8 - (2 - 4 + x) = -4(x - 27 + 3x) - 8$$
$$8 - (-2 + x) = -4(4x - 27) - 8$$
$$8 + 2 - x = -16x + 108 - 8$$
$$10 - x = -16x + 100$$
$$16x - x = 100 - 10$$
$$15x = 90$$
$$x = \frac{90}{15}$$
$$x = 6$$

Check the answer: Plug $x = 6$ into the original equation.

$$8 - [2 - (4 - x)] = -4[x - 3(9 - x)] - 8$$
$$8 - [2 - (4 - 6)] = -4[6 - 3(9 - 6)] - 8$$
$$8 - [2 - (-2)] = -4[6 - 3(3)] - 8$$
$$8 - (2 + 2) = -4(6 - 9) - 8$$
$$8 - 4 = -4(-3) - 8$$
$$4 = 12 - 8$$
$$4 = 4$$

The answer checks out.

Problem 34

Directions: Determine the equation of a straight line that passes through the point $\left(-\sqrt{3}, 4\right)$ and which is perpendicular to the line that passes through the two points indicated below.

$$\left(5, \sqrt{3}\right) \quad \text{and} \quad \left(3, -\sqrt{3}\right)$$

❖ You can find the solution on the following page.

Solution to Problem 34

Find the slope of the line that connects $(x_1, y_1) = (5, \sqrt{3})$ and $(x_2, y_2) = (3, -\sqrt{3})$.

$$m = \frac{y_2 - y_1}{x_2 - x_1} = \frac{(-\sqrt{3}) - \sqrt{3}}{3 - 5} = \frac{-2\sqrt{3}}{-2} = \sqrt{3}$$

We wish to find a new line that is perpendicular to the given line. Recall that lines are perpendicular if one slope is the negative of the reciprocal of the other. That is,

$$m_\perp = -\frac{1}{m}$$

where $m_\perp$ is the slope of the perpendicular line. Plug in $m = \sqrt{3}$.

$$m_\perp = -\frac{1}{\sqrt{3}} = -\frac{1}{\sqrt{3}}\frac{\sqrt{3}}{\sqrt{3}} = -\frac{\sqrt{3}}{3}$$

We multiplied by $\frac{\sqrt{3}}{\sqrt{3}}$ in order to rationalize the denominator. Note that $\sqrt{3}\sqrt{3} = 3$. (If necessary, you can verify with a calculator that $\frac{1}{\sqrt{3}}$ and $\frac{\sqrt{3}}{3}$ both approximately equal 0.577350269.) The equation for the perpendicular line is:

$$y = m_\perp x + b_\perp$$
$$y = -\frac{\sqrt{3}}{3}x + b_\perp$$

To complete our solution, we must find the y-intercept of the perpendicular line. The perpendicular line passes through the point $(-\sqrt{3}, 4)$, so we may plug $x = -\sqrt{3}$ and $y = 4$ into the previous equation and solve for $b_\perp$.

$$4 = -\frac{\sqrt{3}}{3}(-\sqrt{3}) + b_\perp$$
$$4 = \frac{\sqrt{3}\sqrt{3}}{3} + b_\perp$$
$$4 = \frac{3}{3} + b_\perp$$
$$4 = 1 + b_\perp$$
$$3 = b_\perp$$

Our final answer is $y = -\frac{\sqrt{3}}{3}x + 3$.

Check the answer: Graph each line with a graphing calculator or computer.

Problem 35

Directions: Determine the ratio of y to x in the equation below.

$$7x^2 - 6y^2 = -11x^2 - 4y^2$$

❖ You can find the solution on the following page.

Solution to Problem 35

This single equation has two unknowns. Although you can't determine the value of either x or y, you can find the ratio of y to x for the given equation. Combine like terms. Add $6y^2$ and add $11x^2$ to both sides of the equation.

$$7x^2 - 6y^2 = -11x^2 - 4y^2$$
$$7x^2 + 11x^2 = -4y^2 + 6y^2$$
$$18x^2 = 2y^2$$

Divide both sides of the equation by 2.

$$9x^2 = y^2$$

Divide both sides of the equation by x^2.

$$9 = \frac{y^2}{x^2}$$

Squareroot both sides of the equation.

$$\pm 3 = \frac{y}{x}$$

Why $\pm$? That's because $(-3)^2 = 9$ and $(3)^2 = 9$. There are two possibilities: x and y could have the same sign or x and y could have opposite signs. The general solution must allow for both possibilities.

Check the answers: Plug $y = \pm 3x$ into the original equation.

$$7x^2 - 6(\pm 3x)^2 = -11x^2 - 4(\pm 3x)^2$$

Apply the rule that $(ab)^m = a^m b^m$.

$$7x^2 - 6(9x^2) = -11x^2 - 4(9x^2)$$
$$7x^2 - 54x^2 = -11x^2 - 36x^2$$
$$-47x^2 = -47x^2$$

The answers check out.

Problem 36

Directions: Determine what comes next in the following pattern.

$$3x - 2y$$

$$9x^2 - 12xy + 4y^2$$

$$27x^3 - 54x^2y + 36xy^2 - 8y^3$$

❖ You can find the solution on the following page.

Solution to Problem 36

What can you do algebraically to each line in order to create the given pattern?

Multiply the previous line by $(3x - 2y)$. For example, going from the first line to the second line:

$$(3x - 2y)(3x - 2y) = 9x^2 - 6xy - 6xy + 4y^2 = 9x^2 - 12xy + 4y^2$$

Similarly, going from the second line to the third line:

$$(9x^2 - 12xy + 4y^2)(3x - 2y) = 27x^3 - 18x^2y - 36x^2y + 24xy^2 + 12xy^2 - 8y^3$$
$$= 27x^3 - 54x^2y + 36xy^2 - 8y^3$$

The next line in the pattern comes from multiplying $(27x^3 - 54x^2y + 36xy^2 - 8y^3)$ by $(3x - 2y)$.

$$(27x^3 - 54x^2y + 36xy^2 - 8y^3)(3x - 2y)$$
$$= 81x^4 - 54x^3y - 162x^3y + 108x^2y^2 + 108x^2y^2 - 72xy^3 - 24xy^3 + 16y^4$$
$$= 81x^4 - 216x^3y + 216x^2y^2 - 96xy^3 + 16y^4$$

Check the answer: This is actually an application of the binomial expansion.

$$(a + b)^2 = a^2 + 2ab + b^2$$
$$(a + b)^3 = a^3 + 3a^2b + 3ab^2 + b^3$$
$$(a + b)^4 = a^4 + 4a^3b + 6a^2b^2 + 4ab^3 + b^4$$

In our case, a and b are equal to:

$$a = 3x \quad , \quad b = -2y$$

Substitute these expressions into the fourth power of the binomial expansion.

$$(a + b)^4 = a^4 + 4a^3b + 6a^2b^2 + 4ab^3 + b^4$$
$$(3x - 2y)^4 = (3x)^4 + 4(3x)^3(-2y) + 6(3x)^2(-2y)^2 + 4(3x)(-2y)^3 + (-2y)^4$$

Apply the rule that $(ab)^m = a^m b^m$.

$$(3x - 2y)^4 = 81x^4 + 4(27x^3)(-2y) + 6(9x^2)(4y^2) + 4(3x)(-8y^3) + 16y^4$$
$$(3x - 2y)^4 = 81x^4 - 216x^3y + 216x^2y^2 - 96xy^3 + 16y^4$$

This matches the answer that we obtained earlier.

Problem 37

Directions: Apply algebra in order to isolate x in the equation below.

$$\sqrt{a - \sqrt{x}} = b$$

❖ You can find the solution on the following page.

Solution to Problem 37

Square both sides of the equation. Recall that $\left(\sqrt{y}\right)^2 = y$.

$$\left(\sqrt{a - \sqrt{x}}\right)^2 = b^2$$

$$a - \sqrt{x} = b^2$$

Add $\sqrt{x}$ to both sides and subtract b^2 from both sides.

$$a - b^2 = \sqrt{x}$$

Square both sides of the equation. Recall the rule that $(y - z)^2 = y^2 - 2yz + z^2$. In this case, $y = a$ and $z = b^2$.

$$(a - b^2)^2 = \left(\sqrt{x}\right)^2$$

$$a^2 - 2ab^2 + b^4 = x$$

Check the answer: Try it with numbers. (Choose $a > b^2$ in order for the squareroot to yield a real answer.) Let $a = 25$ and $b = 4$. In the equation above, we get:

$$25^2 - 2(25)(4)^2 + 4^4 = x$$

$$625 - 800 + 256 = x$$

$$81 = x$$

See if these numbers satisfy the given equation.

$$\sqrt{25 - \sqrt{81}} = 4$$

$$\sqrt{25 - 9} = 4$$

$$\sqrt{16} = 4$$

$$4 = 4$$

The answer checks out.

Problem 38

Directions: Solve for x and y in the system of equations below, where the numbers with bars over them are repeating decimals. For example, $0.\overline{6} = 0.66666666\cdots$ with the 6 repeating forever and $0.\overline{81} = 0.81818181\cdots$ with the 81 repeating forever.

$$1.\overline{3}x - 0.\overline{81}y = 0.\overline{18}$$

$$0.\overline{6}x + 0.\overline{27}y = 0.\overline{24}$$

❖ You can find the solution on the following page.

Solution to Problem 38

Convert each repeating decimal to a fraction. You can check this with a calculator. (If needed, you may wish to review how to convert repeating decimals into fractions.)

- $1.\bar{3} = 1.33333333\cdots = \frac{4}{3}$. (Why? Because $\frac{1}{3} = 0.33333333\cdots$)
- $0.\overline{81} = 0.81818181\cdots = \frac{9}{11}$. (Why? Because $\frac{81}{99} = 0.81818181\cdots$)
- $0.\overline{18} = 0.18181818\cdots = \frac{2}{11}$. (Why? Because $\frac{18}{99} = 0.18181818\cdots$)
- $0.\bar{6} = 0.66666666\cdots = \frac{2}{3}$. (Why? Because $\frac{6}{9} = 0.66666666\cdots$)
- $0.\overline{27} = 0.27272727\cdots = \frac{3}{11}$. (Why? Because $\frac{27}{99} = 0.27272727\cdots$)
- $0.\overline{24} = 0.24242424\cdots = \frac{8}{33}$. (Why? Because $\frac{24}{99} = 0.24242424\cdots$)

We can rewrite the given equations in terms of fractions as follows:

$$\frac{4}{3}x - \frac{9}{11}y = \frac{2}{11} \quad , \quad \frac{2}{3}x + \frac{3}{11}y = \frac{8}{33}$$

Multiply both equations by 33.

$$44x - 27y = 6 \quad , \quad 22x + 9y = 8$$

Multiply the second equation by 3. Then add the two equations together.

$$44x - 27y = 6 \quad , \quad 66x + 27y = 24$$
$$44x - 27y + 66x + 27y = 6 + 24$$
$$110x = 30$$
$$x = \frac{30}{110} = \frac{3}{11}$$

Plug this into one of the previous equations. We will use $22x + 9y = 8$.

$$22\left(\frac{3}{11}\right) + 9y = 8$$
$$6 + 9y = 8$$
$$y = \frac{2}{9}$$

Check the answers: Plug $x = \frac{3}{11} = 0.\overline{27}$ and $y = \frac{2}{9} = 0.\bar{2}$ into the original equations.

$$(1.33333333)(0.27272727) - (0.81818181)(0.22222222) = 0.18181818$$
$$(0.66666666)(0.27272727) + (0.27272727)(0.22222222) = 0.24242424$$

Use a calculator. The answers check out (apart from a little round-off error).

Problem 39

Directions: Solve for x in the equation below. (Don't use guess and check.)

$$2\sqrt{x} - 14 = \frac{288}{\sqrt{x}}$$

❖ You can find the solution on the following page.

Solution to Problem 39

Multiply both sides of the equation by $\sqrt{x}$. Note that $\sqrt{x}\sqrt{x} = x$ and $\frac{\sqrt{x}}{\sqrt{x}} = 1$.

$$2x - 14\sqrt{x} = 288$$

This is actually a quadratic equation. One way to see this is to let $y = \sqrt{x}$, such that $y^2 = x$.

$$2y^2 - 14y = 288$$

Subtract 288 from both sides of the equation in order to put this in standard form.

$$2y^2 - 14y - 288 = 0$$

Compare this to the general equation $ay^2 + by + c = 0$ to determine that:

$$a = 2 \quad , \quad b = -14 \quad , \quad c = -288$$

Plug these values into the quadratic formula. (Alternatively, you could factor the previous quadratic equation.)

$$y = \frac{-b \pm \sqrt{b^2 - 4ac}}{2a} = \frac{-(-14) \pm \sqrt{(-14)^2 - 4(2)(-288)}}{2(2)} = \frac{14 \pm \sqrt{196 + 2304}}{4}$$

$$y = \frac{14 \pm \sqrt{2500}}{4} = \frac{14 \pm 50}{4}$$

There are two possible answers for y:

$$y = \frac{14 + 50}{4} = \frac{64}{4} = 16 \quad \text{or} \quad y = \frac{14 - 50}{4} = -\frac{36}{4} = -9$$

We're not finished yet because we need to solve for x. Recall that $y = \sqrt{x}$, which equates to $y^2 = x$.

$$x = y^2 = 16^2 = 256 \quad \text{or} \quad x = y^2 = (-9)^2 = 81$$

Check the answers: Plug $x = 81$ and $x = 256$ into the original equation.

$$2\sqrt{256} - 14 = \frac{288}{\sqrt{256}} \quad \text{or} \quad 2\sqrt{81} - 14 = \frac{288}{\sqrt{81}}$$

$$2(16) - 14 = \frac{288}{16} \quad \text{or} \quad 2(-9) - 14 = \frac{288}{-9}$$

$$32 - 14 = 18 \quad \text{or} \quad -18 - 14 = -32$$

The answers check out. However, the solution $x = 81$ only works if you allow for the negative root of $\sqrt{81}$.

Problem 40

Directions: Apply algebra to solve the following word problem.

Two beetles initially separated by a distance of 300 cm walk towards one another at the same constant rate of 2 cm/s. An ant which is initially next to one of the beetles walks away from it, towards the other beetle, at a constant rate of 10 cm/s until it reaches the other beetle. The ant then promptly turns around, traveling with the same constant rate of 10 cm/s until it reaches the previous beetle. This continues until the two beetles both meet the ant in the middle. All together, how far does the ant travel?

❖ You can find the solution on the following page.

Solution to Problem 40

Fortunately, it is completely unnecessary to divide the ant's trip up into pieces and find the distance that the ant travels in each direction before reaching a beetle.

First, determine how much time the ant spends walking. The ant walks until the two beetles meet in the middle. Each beetle travels with a constant speed of 2 cm/s. Each beetle travels a distance of 150 cm (so that the two distances add together to make 300 cm). Divide the distance by the speed (which is a rate) to find the time.

$$t = \frac{d_b}{r_b} = \frac{150}{2} = 75 \text{ s}$$

The ant spends 75 s walking (the same amount of time as the beetles). The ant has a constant speed of 10 cm/s. Multiply the ant's speed (which is a rate) by the time to find the distance that the ant travels.

$$d_a = r_a t = (10)(75) = 750 \text{ cm}$$

The ant travels a total distance of 750 cm (which could also be expressed as 7.50 m, since there are 100 centimeters in one meter).

Check the answer: The ant travels 5 times faster than either beetle (since 10 cm/s is 5 times 2 cm/s). The ant travels for the same amount of time as the beetles. Thus, the ant must travel 5 times farther than either beetle. The ant's distance, 750 cm, is 5 times the distance, 150 cm, that each beetle travels. The answer checks out.

Problem 41

Directions: Solve for x in the equation below. (Don't use guess and check.)

$$(x^2 - x - 6)\{(x^2 - 7x + 12)[(x^2 - 3x - 4)(x + 1)^{-1}]^{-1}\}^{-1} = 9$$

❖ You can find the solution on the following page.

Solution to Problem 41

The given equation has the following structure:
$$p\{q[rs^{-1}]^{-1}\}^{-1} = 9$$
where $p = x^2 - x - 6$, $q = x^2 - 7x + 12$, $r = x^2 - 3x - 4$, and $s = x + 1$. Apply the rule $y^{-1} = \frac{1}{y}$. The above equation is the same thing as:
$$\frac{p}{q \div \frac{r}{s}} = 9$$

Let's begin by looking at just part of the equation.
$$rs^{-1} = \frac{r}{s} = \frac{x^2 - 3x - 4}{x + 1}$$
The numerator can be factored as follows. The denominator cancels.
$$rs^{-1} = \frac{(x - 4)(x + 1)}{x + 1} = x - 4$$

Now let's include the q.
$$q \div \frac{r}{s} = q \div (x - 4) = \frac{q}{x - 4} = \frac{x^2 - 7x + 12}{x - 4}$$
Again we factor the numerator.
$$q \div \frac{r}{s} = \frac{(x - 3)(x - 4)}{x - 4} = x - 3$$

Now include the p.
$$\frac{p}{q \div \frac{r}{s}} = \frac{x^2 - x - 6}{x - 3} = \frac{(x + 2)(x - 3)}{x - 3} = x + 2 = 9$$
$$x = 9 - 2 = 7$$

Check the answer: Plug $x = 7$ into the original equation.
$$(7^2 - 7 - 6)\left\{[7^2 - 7(7) + 12][[7^2 - 3(7) - 4](7 + 1)^{-1}]^{-1}\right\}^{-1} = 9$$
$$36\{12[24(8)^{-1}]^{-1}\}^{-1} = 9$$
$$36\{12[3]^{-1}\}^{-1} = 9$$
$$36\{4\}^{-1} = 9$$
Note that $24(8)^{-1} = \frac{24}{8} = 3$, $12[3]^{-1} = \frac{12}{4} = 3$, and $36\{4\}^{-1} = \frac{36}{4} = 9$. The answer checks out.

Problem 42

Directions: Apply algebra to the two equations given below in order to derive the indicated equation. Note that ω_0 and ω are two distinctly different quantities.

$$\omega = \omega_0 + \alpha\, t$$
$$\theta = \omega_0\, t + \frac{1}{2}\, \alpha\, t^2$$

Use the equations given above to derive the following equation.

$$\omega^2 - \omega_0^2 = 2\, \alpha\, \theta$$

❖ You can find the solution on the following page.

Solution to Problem 42

The equation that we need to derive doesn't involve time (t). Therefore, we should apply algebra to eliminate time from the two given equations. Subtract ω_0 from both sides of the first equation, and then divide by α.

$$\omega - \omega_0 = \alpha t$$

$$\frac{\omega - \omega_0}{\alpha} = t$$

Substitute this expression in place of time in the second equation.

$$\theta = \omega_0 \left(\frac{\omega - \omega_0}{\alpha}\right) + \frac{1}{2}\alpha \left(\frac{\omega - \omega_0}{\alpha}\right)^2$$

Apply the rules $x(y - z) = xy - xz$ and $(y - z)^2 = y^2 - 2yz + z^2$.

$$\theta = \frac{\omega_0\omega - \omega_0^2}{\alpha} + \frac{1}{2}\alpha \left(\frac{\omega^2 - 2\omega_0\omega + \omega_0^2}{\alpha^2}\right)$$

Distribute $\frac{\alpha}{2}$ in the second term. Note that $\frac{\alpha}{\alpha^2} = \frac{1}{\alpha}$.

$$\theta = \frac{\omega_0\omega - \omega_0^2}{\alpha} + \frac{\omega^2 - 2\omega_0\omega + \omega_0^2}{2\alpha}$$

Multiply the first term by $\frac{2}{2}$ in order to make a common denominator.

$$\theta = \frac{2\omega_0\omega - 2\omega_0^2}{2\alpha} + \frac{\omega^2 - 2\omega_0\omega + \omega_0^2}{2\alpha}$$

$$\theta = \frac{2\omega_0\omega - 2\omega_0^2 + \omega^2 - 2\omega_0\omega + \omega_0^2}{2\alpha}$$

The $2\omega_0\omega$'s cancel out. Note that $-2\omega_0^2 + \omega_0^2 = -\omega_0^2$.

$$\theta = \frac{\omega^2 - \omega_0^2}{2\alpha}$$

Multiply both sides of the equation by 2α.

$$2\alpha\theta = \omega^2 - \omega_0^2$$

Check the answer: You can look up the equations of uniform angular acceleration in a standard physics textbook.

$$\omega = \omega_0 + \alpha t \quad , \quad \theta = \omega_0 t + \frac{1}{2}\alpha t^2 \quad , \quad \omega^2 - \omega_0^2 = 2\alpha\theta$$

The answer checks out.

Problem 43

Directions: Find positive solutions for x in the equation below.

$$6x^{3/2} + 15\sqrt{x} = 19x$$

❖ You can find the solution on the following page.

Solution to Problem 43

Note that $\sqrt{x} = x^{1/2}$. Subtract $19x$ from both sides of the equation.

$$6x^{3/2} + 15x^{1/2} = 19x$$

$$6x^{3/2} - 19x + 15x^{1/2} = 0$$

Factor out $x^{1/2}$. Note that $x^{1/2}x = x^{3/2}$ and $x^{1/2}x^{1/2} = x$ (since $x = x^1$).

$$x^{1/2}\left(6x - 19x^{1/2} + 15\right) = 0$$

Either $x^{1/2} = 0$ or $6x - 19x^{1/2} + 15 = 0$. The problem specifically asks for positive solutions, so we will discard $x^{1/2} = 0$.

$$6x - 19x^{1/2} + 15 = 0$$

This is actually a quadratic equation. One way to see this is to let $y = \sqrt{x} = x^{1/2}$, such that $y^2 = x$.

$$6y^2 - 19y + 15 = 0$$

Compare this to the general equation $ay^2 + by + c = 0$ to determine that:

$$a = 6 \quad , \quad b = -19 \quad , \quad c = 15$$

Plug these values into the quadratic formula.

$$y = \frac{-b \pm \sqrt{b^2 - 4ac}}{2a} = \frac{-(-19) \pm \sqrt{(-19)^2 - 4(6)(15)}}{2(6)} = \frac{19 \pm \sqrt{361 - 360}}{12}$$

$$y = \frac{19 \pm \sqrt{1}}{12} = \frac{19 \pm 1}{12}$$

$$y = \frac{19 + 1}{12} = \frac{20}{12} = \frac{20 \div 4}{12 \div 4} = \frac{5}{3} \quad \text{or} \quad y = \frac{19 - 1}{12} = \frac{18}{12} = \frac{18 \div 6}{12 \div 6} = \frac{3}{2}$$

We're not finished yet because we need to solve for x. Recall that $y^2 = x$.

$$x = y^2 = \left(\frac{5}{3}\right)^2 = \frac{25}{9} \quad \text{or} \quad x = y^2 = \left(\frac{3}{2}\right)^2 = \frac{9}{4}$$

Check the answers: Plug $x = \frac{9}{4}$ and $x = \frac{25}{9}$ into the original equation.

$$6\left(\frac{9}{4}\right)^{3/2} + 15\sqrt{\frac{9}{4}} = 19\left(\frac{9}{4}\right) \quad \text{or} \quad 6\left(\frac{25}{9}\right)^{3/2} + 15\sqrt{\frac{25}{9}} = 19\left(\frac{25}{9}\right)$$

$$\frac{81}{4} + \frac{45}{2} = \frac{171}{4} \quad \text{or} \quad \frac{250}{9} + 25 = \frac{475}{9}$$

The answers check out.

Problem 44

Directions: Solve for x and y in the system of equations below.

$$\frac{2}{x} - \frac{3}{y} = 6$$

$$\frac{5}{x} + \frac{8}{y} = 77$$

❖ You can find the solution on the following page.

Solution to Problem 44

The problem is simpler if we make the following definitions.

$$p = \frac{1}{x} \quad , \quad q = \frac{1}{y}$$

In terms of p and q, the given equations become:

$$2p - 3q = 6$$
$$5p + 8q = 77$$

Multiply the first equation by 8 and the second equation by 3.

$$8(2p - 3q = 6) \rightarrow 16p - 24q = 48$$
$$3(5p + 8q = 77) \rightarrow 15p + 24q = 231$$

Add the two equations together. The $24q$ terms cancel out.

$$16p - 24q + 15p + 24q = 48 + 231$$
$$31p = 279$$
$$p = \frac{279}{31} = 9$$

Plug $p = 9$ into one of the previous equations. We will use $5p + 8q = 77$.

$$5(9) + 8q = 77$$
$$45 + 8q = 77$$
$$8q = 32$$
$$q = \frac{32}{8} = 4$$

We're not finished yet because we need to find x and y. Multiply both sides of $p = \frac{1}{x}$ by x to get $px = 1$, then divide by p to get $x = \frac{1}{p}$. Similarly, $y = \frac{1}{q}$.

$$x = \frac{1}{p} = \frac{1}{9} \quad , \quad y = \frac{1}{q} = \frac{1}{4}$$

Check the answers: Plug $x = \frac{1}{9}$ and $y = \frac{1}{4}$ into the original equations.

$$\frac{2}{x} - \frac{3}{y} = \frac{2}{1/9} - \frac{3}{1/4} = 18 - 12 = 6$$
$$\frac{5}{x} + \frac{8}{y} = \frac{5}{1/9} + \frac{8}{1/4} = 48 + 32 = 77$$

The answers check out. Note that $\frac{1}{1/9} = 9$ and $\frac{1}{1/4} = 4$.

Problem 45

Directions: Solve for x in the equation below.

$$\frac{x\sqrt{3}}{3} - 5\sqrt{2} = x\sqrt{3} + 3\sqrt{2}$$

❖ You can find the solution on the following page.

Solution to Problem 45

Subtract $x\sqrt{3}$ from both sides of the equation and add $5\sqrt{2}$ to both sides.

$$\frac{x\sqrt{3}}{3} - x\sqrt{3} = 3\sqrt{2} + 5\sqrt{2}$$

Combine like terms by factoring.

$$x\sqrt{3}\left(\frac{1}{3} - 1\right) = \sqrt{2}(3 + 5)$$

$$x\sqrt{3}\left(\frac{1}{3} - \frac{3}{3}\right) = \sqrt{2}(8)$$

$$x\sqrt{3}\left(-\frac{2}{3}\right) = 8\sqrt{2}$$

$$-\frac{2\sqrt{3}}{3}x = 8\sqrt{2}$$

Multiply both sides of the equation by $-\sqrt{3}$.

$$\frac{2\sqrt{3}\sqrt{3}}{3}x = -8\sqrt{2}\sqrt{3}$$

Note that $\sqrt{3}\sqrt{3} = 3$. Apply the rule $\sqrt{a}\sqrt{b} = \sqrt{ab}$.

$$\frac{2(3)}{3}x = -8\sqrt{(2)(3)}$$

$$2x = -8\sqrt{6}$$

Divide both sides of the equation by 2. The decimal was found using a calculator.

$$x = -4\sqrt{6} \approx -9.797958971$$

Note: If you get $x = -\frac{12\sqrt{2}}{\sqrt{3}} \approx -9.797958971$, your answer is the same (except that you didn't rationalize your denominator, which many math instructors prefer).

Check the answer: Plug $x \approx -9.797958971$ into the original equation.

$$\frac{-9.797958971\sqrt{3}}{3} - 5\sqrt{2} = -9.797958971\sqrt{3} + 3\sqrt{2}$$

$$-5.656854248 - 7.071067812 = -16.97056275 + 4.242640687$$

$$-12.72792206 = -12.72792206$$

Use a calculator. The answer checks out.

Problem 46

Directions: Solve for all possible finite values for x in the equation below. (Don't use guess and check.)

$$(2^{-4} - x^{-3})^{-2} = 256$$

❖ You can find the solution on the following page.

Solution to Problem 46

Apply the rule that $y^{-m} = \frac{1}{y^m}$ where $y = (2^{-4} - x^{-3})$ and $m = 2$.

$$\frac{1}{(2^{-4} - x^{-3})^2} = 256$$

Squareroot both sides. Note that $\sqrt{256} = \pm 16$ since $(-16)^2 = 256$ and $16^2 = 256$.

$$\frac{1}{\sqrt{(2^{-4} - x^{-3})^2}} = \sqrt{256}$$

$$\frac{1}{2^{-4} - x^{-3}} = \pm 16$$

Cross multiply.

$$1 = (2^{-4} - x^{-3})(\pm 16)$$

Divide both sides of the equation by ± 16.

$$\pm \frac{1}{16} = 2^{-4} - x^{-3}$$

Apply the rule that $x^{-n} = \frac{1}{x^n}$.

$$\pm \frac{1}{16} = \frac{1}{2^4} - \frac{1}{x^3}$$

Add $\frac{1}{x^3}$ to both sides, and subtract $\frac{1}{16}$ from both sides. Note that $2^4 = 16$.

$$\frac{1}{x^3} = \frac{1}{16} \mp \frac{1}{16}$$

$$\frac{1}{x^3} = \frac{1}{16} - \frac{1}{16} = 0 \quad \text{or} \quad \frac{1}{x^3} = \frac{1}{16} + \frac{1}{16} = \frac{2}{16} = \frac{1}{8}$$

Since the problem specifically asks for finite answers, we will discard $\frac{1}{x^3} = 0$ as a solution. (Note that the solution to $\frac{1}{x^3} = 0$ is $x = \frac{1}{0}$, which is undefined, as an infinite value of x is needed to solve $\frac{1}{x^3} = 0$.) Invert both sides and take the cube root.

$$\frac{1}{x^3} = \frac{1}{8} \rightarrow x^3 = 8 \rightarrow x = 8^{1/3} = \sqrt[3]{8} = 2 \quad \text{(since } 2^3 = 8\text{)}$$

Check the answer: Plug $x = 2$ into the original equation.

$$(2^{-4} - 2^{-3})^{-2} = \left(\frac{1}{2^4} - \frac{1}{2^3}\right)^{-2} = \left(\frac{1}{16} - \frac{1}{8}\right)^{-2} = \left(\frac{1}{16} - \frac{2}{16}\right)^{-2} = \left(-\frac{1}{16}\right)^{-2} = 256$$

The answer check out. If needed, use a calculator.

Problem 47

Directions: Solve for x in the equation below. (Don't use guess and check.)

$$\frac{30}{x^2} + \frac{21}{x} = -3$$

❖ You can find the solution on the following page.

Solution to Problem 47

Multiply both sides of the equation by x^2. Note that $\frac{x^2}{x^2} = 1$ and $\frac{x^2}{x} = x$.

$$30 + 21x = -3x^2$$

This is a quadratic equation. Add $3x^2$ to both sides of the equation and reorder the terms in order to put it in standard form.

$$3x^2 + 21x + 30 = 0$$

We can factor it, noting that $30 = (5)(6)$ and that $15x + 6x = 21x$.

$$(3x + 6)(x + 5) = 0$$

The two solutions are:

$$3x + 6 = 0 \quad , \quad x + 5 = 0$$
$$3x = -6 \quad , \quad x = -5$$
$$x = -\frac{6}{3} = -2 \quad , \quad x = -5$$

If instead you apply the quadratic formula, you will get the same answers. Compare the equation $3x^2 + 21x + 30 = 0$ to the general form $ax^2 + bx + c = 0$ to see that $a = 3$, $b = 21$, and $c = 30$.

$$x = \frac{-21 \pm \sqrt{21^2 - 4(3)(30)}}{2(3)} = \frac{-21 \pm \sqrt{441 - 360}}{6} = \frac{-21 \pm \sqrt{81}}{6} = \frac{-21 \pm 9}{6}$$

$$x = \frac{-21 + 9}{6} = \frac{-12}{6} = -2 \quad \text{or} \quad x = \frac{-21 - 9}{6} = -\frac{30}{6} = -5$$

Check the answers: Plug $x = -5$ and $x = -2$ into the original equation.

$$\frac{30}{(-5)^2} + \frac{21}{-5} = -3 \quad \text{or} \quad \frac{30}{(-2)^2} + \frac{21}{-2} = -3$$

$$\frac{30}{25} - \frac{21}{5} = -3 \quad \text{or} \quad \frac{30}{4} - \frac{21}{2} = -3$$

$$\frac{30}{25} - \frac{105}{25} = -3 \quad \text{or} \quad \frac{30}{4} - \frac{42}{4} = -3$$

$$\frac{-75}{25} = -3 \quad \text{or} \quad \frac{-12}{4} = -3$$

The answers check out.

Date: _____ Name: _____

Problem 48

Directions: Solve for ✈ and ♥ in the system of equations below using the table that follows.

$$♫\,✈ + 🚌\,♥ = 🔧$$

$$🎁\,✈ - ☆\,♥ = 👄$$

♫	👄	🎁	☆	🚌	🔧
$4☆$	$🚌 + 4$	$2🔧$	2	$♫ - 3$	$👄 ÷ 3$

❖ You can find the solution on the following page.

Solution to Problem 48

First use the table to determine values of the following symbols:

- ✯ = 2 (this was given directly).
- ♪ = 4✯ = 4(2) = 8.
- 🚌 = ♪ − 3 = 8 − 3 = 5.
- 🍩 = 🚌 + 4 = 5 + 4 = 9.
- ✗ = 🍩 ÷ 3 = 9 ÷ 3 = 3.
- 🎁 = 2✗ = 2(3) = 6.

Plug these values into the given equations.

$$8\text{✈} + 5\text{♥} = 3 \quad , \quad 6\text{✈} - 2\text{♥} = 9$$

Multiply the first equation by 2 and the second equation by 5.

$$2(8\text{✈} + 5\text{♥} = 3) \quad , \quad 5(6\text{✈} - 2\text{♥} = 9)$$

$$16\text{✈} + 10\text{♥} = 6 \quad , \quad 30\text{✈} - 10\text{♥} = 45$$

Add the two equations together. The 10♥ terms cancel.

$$46\text{✈} = 51$$

$$\text{✈} = \frac{51}{46} \approx 1.108695652$$

Plug this value into one of the previous equations. We will use $8\text{✈} + 5\text{♥} = 3$.

$$8\left(\frac{51}{46}\right) + 5\text{♥} = 3$$

$$\frac{408}{46} + 5\text{♥} = 3$$

$$\frac{204}{23} + 5\text{♥} = 3$$

$$5\text{♥} = 3 - \frac{204}{23} = \frac{69}{23} - \frac{204}{23} = -\frac{135}{23}$$

$$\text{♥} = -\frac{135}{23(5)} = -\frac{27}{23} \approx -1.173913043$$

Check the answers: Plug ✈ and ♥ into the original equations. Use a calculator.

$$8\text{✈} + 5\text{♥} \approx 8(1.108695652) + 5(-1.173913043) \approx 3$$

$$6\text{✈} - 2\text{♥} \approx 6(1.108695652) - 2(-1.173913043) \approx 9$$

The answers check out (to within reasonable rounding tolerance).

Problem 49

Directions: Apply algebra to solve the following word problem.

If a roller coaster begins from rest at a height h above the bottom of a circular loop, then at the top of the loop neither the passengers nor their belongings will fall out of the roller coaster if the following inequality holds, where R is the radius of the loop. Rewrite the inequality in terms of the diameter (instead of the radius).

$$h > \frac{5R}{2}$$

❖ You can find the solution on the following page.

Solution to Problem 49

This should be an easy problem, yet there is a common intuitive mistake that many students make. Most students realize that a factor of 2 is involved, since diameter is twice as large as the radius. However, that factor of 2 is often applied incorrectly.

If, as the problem instructs, you force yourself to express your ideas with algebraic equations and you make careful substitutions, it greatly increases the chances that you will solve the problem correctly. (If you answered this problem correctly by any means, pat yourself on the back. If not, don't feel bad: It's common; try to learn from your mistake.)

Diameter (D) equals twice the radius (R):
$$D = 2R$$
The given inequality has radius, which we need to eliminate. Therefore, we need to solve for radius in the equation above. Divide both sides of the equation by 2.
$$R = \frac{D}{2}$$
Plug this expression in place of radius in the given inequality.
$$h > \frac{5R}{2} = \frac{5}{2}R = \frac{5}{2}\left(\frac{D}{2}\right) = \frac{5D}{4}$$

Check the answer: Let's try a number for radius, like $R = 40$ m, and see if our two inequalities are consistent. In the original inequality, we get:
$$h > \frac{5R}{2} = \frac{5(40)}{2} = \frac{200}{2} = 100 \text{ m}$$
The diameter is twice the radius:
$$D = 2R = 2(40) = 80 \text{ m}$$
In the derived inequality, we get:
$$h > \frac{5D}{4} = \frac{5(80)}{4} = \frac{400}{4} = 100 \text{ m}$$
You can see that we obtained the same minimum height (100 m) both ways. The answer checks out.

Problem 50

Directions: Solve for x in the system of equations below.

$$xy = z\sqrt{3}$$

$$y\sqrt{1 - x^2} = z$$

❖You can find the solution on the following page.

Solution to Problem 50

There are three variables in this equation: x, y, and z. Ordinarily, you would need three independent equations in order to solve for three unknowns. However, in this case, you can solve for x in the given equation. The trick is not to give up before you try. We can eliminate both y and z by dividing the two equations. (Can you divide two equations? Yes. If $a = b$ and $c = d$, then it must be true that $\frac{a}{c} = \frac{b}{d}$, provided that we're not dividing by zero.)

$$\frac{xy}{y\sqrt{1-x^2}} = \frac{z\sqrt{3}}{z}$$

Note that $\frac{y}{y} = 1$ and $\frac{z}{z} = 1$.

$$\frac{x}{\sqrt{1-x^2}} = \sqrt{3}$$

Square both sides of the equation. Apply the rule that $\left(\sqrt{y}\right)^2 = y$.

$$\frac{x^2}{1-x^2} = 3$$

Multiply both sides of the equation by $1 - x^2$.

$$x^2 = 3(1 - x^2)$$
$$x^2 = 3 - 3x^2$$

Add $3x^2$ to both sides of the equation.

$$4x^2 = 3$$

Divide both sides of the equation by 4.

$$x^2 = \frac{3}{4}$$

Squareroot both sides of the equation. The decimal was found using a calculator.

$$x = \sqrt{\frac{3}{4}} = \frac{\sqrt{3}}{\sqrt{4}} = \frac{\sqrt{3}}{2} \approx 0.866025404$$

Check the answer: Plug $x \approx 0.866025404$ into the original equations.

$$0.866025404y = z\sqrt{3} \rightarrow y = z\sqrt{3}/0.866025404 \rightarrow y = 2z$$
$$y\sqrt{1 - 0.866025404^2} = z \rightarrow y\sqrt{1 - 0.75} = z \rightarrow y\sqrt{0.25} = z \rightarrow 0.5y = z \rightarrow y = 2z$$

The answer checks out (since both equations lead to $y = 2z$).

WAS THIS BOOK HELPFUL?

A great deal of effort and thought was put into this book, such as:

- Careful selection of problems for their instructional value.
- Breaking down the solutions to help make the math easier to understand.
- Coming up with a good variety of ways to offer a challenge.
- Multiple stages of proofreading, editing, and formatting.
- Beta testers provided valuable feedback.

If you appreciate any of the effort that went into making this book possible, there is a simple way that you could show it:

<u>Please take a moment to post an honest review.</u>

For example, you can review this book at Amazon.com or Barnes & Noble's website at BN.com.

Even a short review can be helpful and will be much appreciated. If you're not sure what to write, following are a few ideas to help you get started, though it's best to describe what is important to you.

- Did you enjoy the selection of problems?
- Were you able to understand the solutions and explanations?
- Do you appreciate the handy formulas on the back cover of the print edition?
- How much did you learn from reading and using this workbook?
- Would you recommend this book to others? If so, why?

Do you believe that you found a mistake? Please email the author, Chris McMullen, at greekphysics@yahoo.com to ask about it. One of two things will happen:

- You might discover that it wasn't a mistake after all and learn why.
- You might find out that you're right, in which case the author will be grateful and future readers will benefit from the correction. Everyone is human.

ABOUT THE AUTHOR

Dr. Chris McMullen has over 20 years of experience teaching university physics in California, Oklahoma, Pennsylvania, and Louisiana. Dr. McMullen is also an author of math and science workbooks. Whether in the classroom or as a writer, Dr. McMullen loves sharing knowledge and the art of motivating and engaging students.

The author earned his Ph.D. in phenomenological high-energy physics (particle physics) from Oklahoma State University in 2002. Originally from California, Chris McMullen earned his Master's degree from California State University, Northridge, where his thesis was in the field of electron spin resonance.

As a physics teacher, Dr. McMullen observed that many students lack fluency in fundamental math skills. In an effort to help students of all ages and levels master basic math skills, he published a series of math workbooks on arithmetic, fractions, and algebra entitled *Improve Your Math Fluency*. Dr. McMullen has also published a variety of science books, including introductions to basic astronomy and chemistry concepts in addition to physics workbooks.

Author, Chris McMullen, Ph.D.

ALGEBRA

For students who need to improve their algebra skills:
- Isolating the unknown
- Quadratic equations
- Factoring
- Cross multiplying
- Systems of equations
- Straight line graphs

www.improveyourmathfluency.com

SCIENCE

Dr. McMullen has published a variety of **science** books, including:

- Basic astronomy concepts
- Basic chemistry concepts
- Balancing chemical reactions
- Calculus-based physics textbooks
- Calculus-based physics workbooks
- Calculus-based physics examples
- Trig-based physics workbooks
- Trig-based physics examples
- Creative physics problems

www.monkeyphysicsblog.wordpress.com

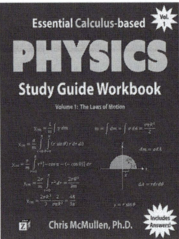
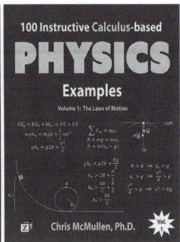

MATH

This series of math workbooks is geared toward practicing essential math skills:

- Algebra and trigonometry
- Fractions, decimals, and percentages
- Long division
- Multiplication and division
- Addition and subtraction

www.improveyourmathfluency.com

PUZZLES

The author of this book, Chris McMullen, enjoys solving puzzles. His favorite puzzle is Kakuro (kind of like a cross between crossword puzzles and Sudoku). He once taught a three-week summer course on puzzles. If you enjoy mathematical pattern puzzles, you might appreciate:

300+ Mathematical Pattern Puzzles

Number Pattern Recognition & Reasoning
- pattern recognition
- visual discrimination
- analytical skills
- logic and reasoning
- analogies
- mathematics

VErBAl ReAcTiONS

Chris McMullen has coauthored several word scramble books. This includes a cool idea called **VErBAl ReAcTiONS**. A VErBAl ReAcTiON expresses word scrambles so that they look like chemical reactions. Here is an example:

$$2\,C + U + 2\,S + Es \rightarrow S\,U\,C\,C\,Es\,S$$

The left side of the reaction indicates that the answer has 2 C's, 1 U, 2 S's, and 1 Es. Rearrange CCUSSEs to form SUCCEsS.

Each answer to a **VErBAl ReAcTiON** is not merely a word: It is a chemical word. A chemical word is made up not of letters, but of elements of the periodic table. In this case, SUCCEsS is made up of sulfur (S), uranium (U), carbon (C), and Einsteinium (Es).

Another example of a chemical word is GeNiUS. It's made up of germanium (Ge), nickel (Ni), uranium (U), and sulfur (S).

If you enjoy anagrams and like science or math, these puzzles are tailor-made for you.

BALANCING CHEMICAL REACTIONS

$$2\,C_2H_6 + 7\,O_2 \rightarrow 4\,CO_2 + 6\,H_2O$$

Balancing chemical reactions isn't just chemistry practice.

These are also **fun puzzles** for math and science lovers.

Balancing Chemical Equations Worksheets
Over 200 Reactions to Balance
Chemistry Essentials Practice Workbook with Answers
Chris McMullen, Ph.D.

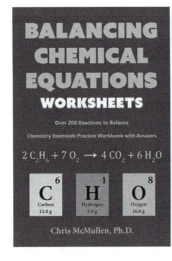

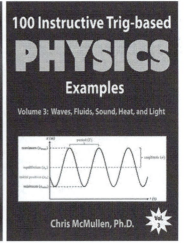